Interpretation and Tour Leadership

Principles and Practices of Tour Guiding

Nimit Chowdhary and Monika Prakash

(G) Goodfellow Publishers Ltd

(G) Published by Goodfellow Publishers Limited,
26 Home Close, Wolvercote, Oxford OX2 8PS
http://www.goodfellowpublishers.com

British Library Cataloguing in Publication Data: a catalogue record for this title is available from the British Library.

Library of Congress Catalog Card Number: on file.

ISBN: 978-1-911635-95-6

DOI: 10.23912/9781911635956-4986

Design and typesetting by P.K. McBride, www.macbride.org.uk

Cover design by Cylinder

Printed by Printforce, Biggleswade

Distributed by UK Marston Book Services, www.marston.co.uk

Contents

About the authors

Prof. Nimit Chowdhary

Dr Nimit Chowdhary is an engineer with, MBA and PhD in Management. He is a professor and ex-Head of the Department of Tourism and Hospitality Management at Jamia Millia Islamia (a Central University) who has more than 28 years of academic experience. He has been a full professor for 15 years. He has taught at Indian Institute of Tourism and Travel Management (Gwalior, Noida, Nellore), Mizoram University (Aizawl), Jiwaji University (Gwalior), Maharshi Dayanand Saraswati University (Ajmer), Instituto Tecnológico y de Estudios Superiores de Monterrey (México) and Shaoxing University, (P.R. China). He has also taught in Sweden, Spain and Slovenia.

He is a recipient of the AICTE Career Award for Young Teachers; SIDA Fellowship, Sweden; Guest Scholarship, Sweden; Linnaeus Palme Exchange Programme Grants, Sweden; PIMG Research Excellence Award, Gwalior; Scholars' Grant (EMTM), Erasmus Mundus, Europe, among others. He was also chosen for the prestigious LEAP programme at Oxford University. He has supervised 15 PhDs, authored eight books, edited five books and contributed more than 155 papers.

Prof. Monika Prakash

Dr Monika Prakash is a Professor at the North India Campus of Indian Institute of Tourism and Travel Management (IITTM), Noida, with the additional charge of Head of the South India Campus of the institute. She is currently leading the prestigious Incredible India Tourist Facilitator Programme (IITFP) of the Ministry of Tourism, Government of India. In addition, she is a researcher, trainer, and academic administrator.

She has a double master's in Commerce (MCom) and Management (MBA) and a PhD in Commerce.

She has published more than 45 papers and eight books. She has supervised doctoral research and led funded research from the Ministry of Tourism and ICSSR. She has presented papers and moderated/ chaired national and international conferences sessions. She has travelled extensively in and outside of the country.

Preface

This work stands on our more than total 30 years of experience of engaging with tour guides as trainers. We have had the opportunity to engage with the various governments that wanted to roll out training programmes for tour guides as a capacity-building initiative. However, each agency had a different perspective that needed to be included in the design of training programmes. As a result, we trained more than 4500 tour guides and spent time discussing the challenges they faced. These multiple perspectives enriched our understanding of the tour guiding phenomenon. The first author also supervised two doctoral research studies and published some research on tour guide training during this period. Looking to the need of trainees, the first author wrote the *Handbook for Tour Guides* in 2013, which is more like a training manual. This book deals with various tour guiding situations and practicalities like handling demanding customers or difficult situations. In addition, the book presents for the first time the dynamics and mechanics of tour guiding that bring to fore the essential communication techniques.

Along with other colleagues, we also engaged with tour operators and DMCs to train tour managers, most of whom were licensed tour guides in India. Their role involved enlarged responsibilities of logistics and service. However, not many readings were available to support this training which necessitated another book, *Tour Leadership and Management*, in 2018 with Dr Shailja Sharma. This project was also an opportunity to reflect on the earlier book, which was clinical but had little theory to explain the underlying phenomenon. With the vocationalisation of education, tour guiding has become a significant job role in tourism economies worldwide. As a result, many university and college programmes have introduced tour guiding as a course in their curriculum. They require textbooks. Several books are available, in any case.

Freeman Tilden (1883-1980) pioneered modern heritage interpretation. He worked with the United States National Park Service and collated his understanding in the book *Interpreting Our Heritage* in 1957, one of the discipline's most authoritative texts. He shaped the tour guiding profession by articulating his famous six principles of interpretation which suggested that interpretation is an art and not just sharing of information. Moreover, the primary purpose of interpretation is not instruction but provocation. He famously quoted, "Through interpretation, understanding; through understanding, appreciation; through appreciation, protection." In 1990, Marc Mancini made an early attempt to author a book that comprehensively covers the subject. He authored *Conducting Tours: A Practical Guide*. Divided into ten chapters, it begins with valuable information about guiding, escorting, and tour planning. Next, it discusses subtler topics such as guide commentary, 'tricks' for effective touring, psychology of group behaviour, successful itineraries, and ethical challenges presented to tour leaders.

Another pioneer was Sam H Ham of the University of Idaho, who extended Professor William J. Lewis' (University of Vermont) concept of thematic interpretation to heritage interpretation. He formally presented the thematic approach in his 1992 book *Interpretation: Making a Difference on Purpose,*. He introduced the famous TORE (Thematic, Organised, Relevant and Enjoyable) model. Since then, this book has been standard reading for many interpretations and interpretive tours guiding students.

However, the most popular book on tour guiding used like a textbook is Kathleen Lingle Pond's 1993 book *The Professional Guide: Dynamics of Tour Guiding*. This book, divided into 13 chapters, is more like a reference reading discussing the profession, role, and tour guiding techniques. It also dedicates discussion to the management of the tour guiding business. In 1998 Larry Beck authored *Interpretation for the 21st Century: Fifteen Guiding Principles for Interpreting Nature and Culture*. A revised edition was published in 2002. The book attempts to crystalise Tilden's principles in a handy format by highlighting 15 principles, like lighting a spark, interpreting in the information age, the importance of the story, provocation, holistic interpretation, passion, and modern tools of interpretation, among others.

As research on tour guiding became commonplace, Betty Weiler and Rosemary Black, in 2014, authored *Tour Guiding Research: Insights, Issues and Implications*. The book reviews tour guiding scholarship and research over the previous fifty years. This cogent and concise desk reference book fosters best practice and stimulates further study and research on tour guiding across various disciplines. In addition, it explores how tour-guiding theory and practice has evolved and what factors have contributed to this.

In 2020, Gulsun Yildirim, Ozlem Ozbek, Ceyhun Caglar Kilinc and Abdullah Tarinc edited *Cases on Tour Guide Practices for Alternative Tourism*. This work analyses emerging research exploring the theoretical and practical aspects of tour guides' occupational issues and their conduct in international tourism like cultural education, specialised learning, and international business. This recently published book deals with issues like branding, cultural education, gender inequality, specialised learning, sustainability, and tour guidance, among other topics.

For management of tours, *Handbook of Professional Tour Management* by Robert T. Reilly was published first in 1982. It provided professional literature covering all tour conducting and tour guiding by tour managers. A good handbook for novice tour managers, it deals with planning, developing, and costing a tour, emphasising how to organise a tour from beginning to end.

Interpretation and Tour Leadership: Principles and Practices of Tour Guiding is positioned as a textbook to back teaching in tourism programmes. The book is intended to have a solid theoretical base. However, it will also be helpful for training programmes. While tour-guiding and tour-leadership are clinical courses, techniques need to be backed by relevant theory. Therefore, while most chapters have a sound theoretical base, some focus on transferable techniques and skills.

Like other vocations, tour guiding has also undergone many changes. Impacts of technology must be reckoned. Similarly, the Covid-19 pandemic will have far-reaching impacts on the profession. Thus, a fresh look at the profession of tour guiding is imperative. This book focuses on the job of a tour guide whose role is often expanded to include the responsibilities of a tour escort. Often these are referred to as tour leaders. The terms tour guide, tour manager, tour escort and tour leaders are different but used interchangeably in practice and different contexts. The difference is highlighted in one of the discussions. This work, therefore, touches two dimensions of the role: one, interpretation and guiding techniques (strictly for tour guides); and two, aspects of tour management that the tour leader must understand and possess the necessary skills like handling transfers, leading on day tours, and even multiple day tours where the tour leader needs to liaison with the hotel, and attractions, among others.

The role of the tour guide is clinical. The book's purpose is to back practice with the theory so that the learners have the necessary theoretical foundation for the concepts related to tour guiding. The book begins by introducing a tour guide in Chapter 1. First, the importance of tour guides in a tourism context is established. Next, it is emphasised that visitors require a mediator in an unknown destination who is invariably the tour guide. Subsequently, a brief discussion on the historical evolution of a tour guide role is presented to set a premise. A critical analysis follows this discussion to put the seemingly similar roles in perspective. Expectations of guests and other stakeholders from tour guides are the basis for crystalising the guide's job description and job specification. The chapter also brings to light the role of technology in tour guiding. Finally, it deliberates on how tour guides make money.

After introducing the need, importance, and role of a tour guide, it is essential to discuss the most important stakeholder. Chapter 2 discusses the tourists. A tour guides role envisages spending much time with the visitors and servicing their expectations. In addition, tourists in different stages of travel require support from guides. Most significantly, they are in a different cultural setting and seek guidance to negotiate. There is a discussion on responsible guiding, which deals with managing visitor impacts – environmental, socio-cultural, and economic. Guides also need to remain aware of the requirements of the limited mobility guests. These visitors are a growing segment as tourism intends to turn more inclusive.

Tour guiding is essentially communication between the guide and the visitor. Chapter 3 lays the conceptual foundation for a meaningful discussion on communication and how tour guides need to consider and prepare for each stage of the process. Another crucial thing that tour guides must understand is that the visitors are often from different cultural settings, and cultural difference plays a vital role in cross-cultural communication – both verbal and nonverbal. The heart of this book is Chapter 4 that discusses the art of interpretation. It references Tilden's principles and Sam Ham's TORE model to offer practical tips to tour guides. One of the contributions of this chapter is identifying the tools and techniques that establish the TORE principles. These include ones that we have used training tour guides. Finally, it offers practical tips for tour guiding.

A tour guide is conceptually ready for action with a basic understanding of communication and interpretation. However, on-ground tour guiding needs much planning. Chapter 5 discusses planning for interpretation by a tour guide. It rolls out the planning required by a guide. It emphasises the need for proper research to identify interpretation goals and subsequent message design for the targeted audience. A detailed itinerary is the product of such an exercise. The product also includes customary introductions and briefings. Planning also entails anticipating and addressing visitor concerns. The chapter closes with a discussion on planning a tour guide career.

Having discussed the theoretical underpinnings for interpretation planning, interpretation and interpretive narratives, Chapter 6 presents action: how the tour guides should guide visitors. In the basic format, a tour guide accompanies guests around a monument or a natural site. At a practical level, the chapter details how to walk and talk. This chapter discusses the dos and don'ts of guiding. It also details guiding on a moving vehicle while a section is devoted to the tools that a guide would require while guiding. If Chapter 4 was the heart of the book, Chapter 7 is the soul, and details how to guide tourists in different scenarios – cultural and natural. The chapter discusses how to guide at an archaeological site, religious site, museum, or nature walk.

After discussing the nuances of tour guiding, Chapter 8 focuses on professional challenges faced by tour guides. It elaborates on handling demanding guests and difficult situations. One of the critical skills required by tour guides is to handle questions. This chapter explains this in detail. Additionally, this chapter discusses leading guests in a post Covid scenario. A limiting case is leading a group of visitors on multi-day tours. Guide also have the onus of leading the group responsibly.

Finally, Chapter 9 is about institutionalising the tour guiding business. It presents the various options for creating a firm. It also discusses the important stakeholders and how a tour guide may associate with them. The chapter briefly sheds light on the relevant legal framework for guiding and ethical conduct by a tour guide.

Tour guiding involves much storytelling, so this book is dotted with examples, vignettes and stories. Each chapter begins with clearly delineating the learning objectives. Then, the text is presented in a free-flowing easy to understand language. Finally, the central learning is summarised as Key Points at the end of the chapter. Keywords that came up for discussion are also revisited at the end. In addition, there are chapter-end discussion questions and activities that are intended to help learners reflect on what was discussed in the chapter.

It is also pertinent to mention that we have used the term 'tour guide' more often in the book. However, the term 'tourist guide' is popular in Europe and some other parts of the world. The two are synonymous.

As agreed with the publisher, the chapters of this book will also be available independently to readers. Thus one can customise to address their learning needs. While the text is intended as essential reading for a course on tour guiding, a combination of chapters can be used as training material for tour guide training programmes.

Any academic work is, at best, a work in progress, and there is always room for improvement. We concede that. While we have made all efforts to make this work worthwhile, we seek feedback from readers and friends to improve it further. The same would be received with gratitude. This effort is just the beginning of a long journey to improving tour guiding and interpretation qualitatively.

Happy learning!

Prof Nimit Chowdhary
Jamia Millia Islamia, New Delhi

Prof. Monika Prakash
Indian Institute of Tourism and Travel Management, Noida

Acknowledgements

We have always believed that any work of some significance is not possible without the support of known and unknown noble influences. The Indian tradition also mandates that there is no penance for ingratitude. Therefore, it is incumbent upon me to acknowledge all such stimuli, inspirations, and backing.

First and foremost, we must acknowledge the works of many great predecessors like Freeman Tilden and Sam Ham, KL Pond, Erik Cohen, Betty Weiler and her associate researchers and the like for laying the foundation for discussion on tourist guiding. Next, we acknowledge the tour guides and managers, those foot soldiers of the trade who have suffered the most during the Covid pandemic. We have had the opportunity to train more than three thousand tour guides and managers over the last 15 years, which returned those valuable insights into this vocation and helped a great deal when we write on the subject, including this work. Many of these professionals, often referred to as Regional Level Guides (RLGs) in India, are friends and swiftly answer our queries about the trade. Further, on occasions, they have reviewed our thoughts and notes.

We must also record our appreciation for colleagues from our organisations, Jamia Millia Islamia and the Indian Institute of Tourism and Travel Management. Ramakrishna, Dr Ramesh Devrath and Dr Shailja Sharma from IITTM – the last two being our scholars, colleagues, and co-trainers in many of the training programmes that we organised on campus and outdoors. These training programmes on the ground have allowed us to interact with numerous tour guides at different stages of their careers and understand their concerns. Their inputs came in handy as we worked to make this work more clinical.

We acknowledge our considerate colleagues who bore with us at our present organisations as we devoted time to this and other academic pursuits. When the families went into a huddle during Covid restrictions, we stole some family time to be on the table reading, researching, and writing. Our parents and sons deserve an apology from us.

We must also thank our team, especially scholars at Jamia Millia Islamia, who helped us with reviews, revisions, and editing. Rohan, Pinaz, Ashraf, Khaled, Arman and the entire team deserve mention. Snigdha deserves a special mention for helping with numerous rounds of manuscript reading. RLG colleagues Rajesh PR and Govind Agarwal quickly responded to our queries.

Finally, we acknowledge with gratitude the support from the publishing team, especially Sally North, for bearing with us when it got challenging to work on the manuscript due to Covid 19.

Prof Nimit Chowdhary, Prof Monika Prakash

Dedication

To Anuj Saxena and Vineet Saxena, whom we lost in 2021, and the brethren in the tourism fraternity who faced countless hardships during the Covid 19 pandemic.

1 Introduction to Tour Guiding

Learning objectives

After studying this chapter, you should be able to understand:

- That the importance and nature of tourism mandate a role for tour guides.
- That conducted tours require mediation.
- The historical evolution of the tour guiding vocation.
- The jargon used in guiding and interpretation, and appreciating the different roles.
- Guest expectations from tour guides, their job description and thus the typical job specification.
- How technology impacts tourist guiding and tour guides.
- How tour guides are compensated.

Neeraj felt deceived and at a loss as he relaxed in the hotel lobby and talked to other guests after a long day visit to the famous Chittorgarh Fort, a UNESCO World Heritage site. Neeraj is a travel enthusiast and reads a lot about tourist places and visits them with his family, especially with his young son. His reading comes in handy as he shares his understanding of the place visited with the family. They had spent half a day at the fort. As he sat in the lobby, guests from another tour party sat there animatedly discussing their exciting visit to a small Village Nagari on their tour guide's advice. It was 12 km from the fort and had an archaeological park with a *Hathi-bada* complex where an inscription was found dated 300BCE. The inscription confirms the building of *Puja-Sila-Prakara*– a compound around the stone that's worth veneration. Thus, it is the oldest available Hindu temple compound. Any tourist travelling this far would never like to skip such an important cultural attraction which Neeraj had missed. The other party's guide had enriched their experience by improvising the standard itinerary and adding such marvels.

He recalled how a guide on their previous visit to Gwalior had helped them decipher the place. What appeared to be a creative pattern on stone was actually a *Shanmukha*. Placed on the southern wall of Man Mandir of Mansingh Palace,

Gwalior, it is a six-faced icon which is a union of two triangles – while one represents the *Shiva*, the inverted triangle represents *Shakti*. The combination of masculine and feminine generates a six-faced *Kartikeya* (*Shanmukha*), the progeny of *Shiva* and *Shakti*, and was a good way of presenting the "Creation of Life" to the public.

Tour guides add meaning to what visitors see and do.

Figure 1.1: A Shanmukha on the wall of Gwalior Fort. Picture credit: Ramakrishna Kongalla

Introduction

Travel has become all the more important. The world has changed post-Covid-19 for us. Both internally and externally, we face a new world. Covid has forced individuals to stop and contemplate the issues of existence and our connection to this world. As a result, people look forward to travelling more, knowing more about the world and experiencing the various ways in which animate elements connect to the inanimate. Tourism has become a part of a lifestyle for people. Therefore, tourism is today one of the priorities of society.

The concept of tourism has phenomenally developed over the past four decades. It is being recognised as a vital economic force that has a significant potential to change the condition of a country or a destination. We are living in an experiential economy. Many activities and services nowadays aim to involve the consumer in some kind of live-through. People want to have new experiences, try different things, experiment with different ideas, and coexist in different possible ways. They travel for this reason. Everyone is a 'tourist' at one point or the other. Today, travelling has become a way of life for many. People are travelling, cutting across geographical and political boundaries. However, the reasons for travelling may be different. When people have more disposable time and money available, they try new ways to spend them. Travelling to many seems to be the obvious choice.

The term tourism has many connotations. To some, it may mean travelling for fun during leisure while for others it may mean holidaying, which involves some excitement, adventure, thrill, and challenge. For many, it is a journey towards the realisation of dreams. It is like one has yearned for years to be at their dream destination. Whatever it may be, the thrill of 'unknown' gives a kick to a person when they travel. The wholesome 'experience' that one wants to undergo while visiting a destination immerse themselves entirely into this feeling of thrill and fun, and later share their experiences with friends and family. Though it might sound like a non-serious-leisure activity, it has been of late recognised as an important economic activity. Tourism today is a dynamic industry that requires the contribution and cooperation of other sectors and stakeholders.

Increasingly the benefits of tourism and its contribution towards the economy are being realised; tourism is being recognised as an industry with a lot to offer in the socio-economic development of the destination. Its importance was recognised in the Manila Declaration on World Tourism of 1980 as "an activity essential to the life of nations because of its direct effects on the social, cultural, educational, and economic sectors of national societies and their international relations." For many countries worldwide, tourism is the mainstay of their economy. Like other industries, tourism requires skilled professional human resources to manage the various resources and deliver the required services.

Tourism is travel for recreational, leisure, or business purposes, usually for a limited duration. Tourism is commonly associated with trans-national travel but may also refer to travel to another location within the same country. The World Tourism Organization (UNWTO) defines tourists as people travelling to and staying in places outside their usual environment for not more than one consecutive year for leisure, business and other purposes. Tourism today is an industry managed by professionals who work for a range of enterprises, from large corporations to one-person companies, from for-profit companies to large community initiatives.

Travel is a basic human need

It is not that people have started travelling now. Humans have always travelled. As per the accepted view, modern humans originated somewhere in southern Africa, some 250,000 to 300,000 years ago. Since then, man has travelled to all corners of the world. They travelled for a variety of reasons.

A great trade route originated in the second century BCE, which linked China with the Roman Empire. It lasted till the end of the fourteenth century AD. The route originated from Chang'an (now Xian) in the east and ended at the Mediterranean in the West. This route brought together the peoples and cultures of China, India, Persia, Arabia, Greek and Rome. Besides trade, this route introduced the religions of the West and the incredible inventions of China to other civilisations of the time (Sharma et al., 2014).

The history of civilisation in India is also the history of exchange between peoples of the world, with India being a key destination since times immemorial. It is coetaneous with Chinese, Mesopotamian, Mayan, Egyptian, Greek and Roman civilisations. There are shreds of evidence of receiving visitors – scholars, ambassadors, business people, invaders and casual travellers. These visitors have come from both East and West. As a seat of learning and spiritualism, India is also believed to have sent out emissaries, especially those who have taken Buddhism far and wide. One of the earliest visitors to India for trade were the Jews. They came to India in different waves and for different reasons but primarily for trade. The traditional account is that traders from Judea arrived in Cochin, Kerala, in 562 BCE and that more Jews came as exiles from Israel in 70 AD after the destruction of the Second Temple. According to Indian Christian traditions, the apostle Thomas arrived in Kodungallur (also Muziri) Kerala, in 52 AD, established the seven churches and evangelised in present-day Kerala and Tamil Nadu. It is believed that Christianity was introduced to India to spread the gospel amongst Kerala's Jewish settlements (Sharma et al., 2014).

With technology, travel gained impetus, but this is perhaps not the only reason tourism has grown over the past few decades. Humans have always travelled distances. In the ninth and tenth centuries AD, Vikings occupied some of the most challenging terrains throughout Europe and North America. Pilgrimages have been part of most traditions of the world. When setting out on journeys, people like Columbus, Marco Polo and Vasco da Gama were often unsure if they would ever get back to their homes. Still, they travelled far and wide, searched new lands and new people.

Humans have always been curious and are always excited by the ideas of finding new ways of life, new people, new cultures, and examples of how a man can exist in different ways. A famous Chinese proverb says *`travelling 100 miles is like reading a good book '*. Fa Xian and I-Tsing braved all odds to visit Nalanda to learn. Exposure to new ways of life has always ignited a curiosity and contributed to human creativity and building a world view. Most cultures, especially those that have progressed, have in one way or other promoted travel and through travel exchange. Some have travelled for learning, others for business and many for pilgrimage. Whatever be the reason, man has always enjoyed travelling. However, travelling has always had associated risks and concerns. There is always a fear related to the unknown though this unknown has its share of excitement associated with travel. The business of tourism is mostly about managing this unknown and the risks associated with it.

When a person travels to a different place, he is often excited to see how people in different places and cultures have a world of their own and relate to things around them. This learning is a thrill and is often exciting. One school of thought suggests that tourism is about 'ritual inversion', which means that people travel to see and make sense of things in a place that is different, almost 'inverted', from theirs. People from beaches would love to go to deserts and vice versa; from plains, they will go to hills and vice versa; from crowded cities, they would prefer to go to quiet, sparsely populated places, and vice versa. It is not only about going to a very different place; the fun is about making sense of things at that destination – how people live, why they live that way, why they had been living that way. This learning, in simple terms, is education. People educate themselves about different cultures and places. Learning and knowing is basic human need. A person feels suffocated if he is cut from the world around him. If his information channels are blocked, he instantly stops making sense of things and feels a 'disconnect'. This disconnect could be depressing. This disconnect is why many do not adjust to different cultures and are uncomfortable in unfamiliar places. They fail to establish a rapport and therefore are not able to make sense of things around them. They feel out of place and alienated and are dissatisfied.

On the other hand, if they start learning about a different place and making sense, they are excited and want to know more to be internally comfortable. This education becomes entertaining, and tourism is, therefore, '*edutainment*'. Tourism is about life education. Education, in turn, can be understood at two levels. First, at the elementary level, education is to know about people and places. This education is factual information. This information is available in books, on TV and now through the internet on one's smartphone. One can also know about places and people by hearing from others. The second and higher level of learning is about realisation. The realisation comes through being immersed in a situation.

Search for *Taj Mahal* on Google. Millions of links are returned. The Taj Mahal's excellent views and moods have been captured and posted on the internet. You can have a full moon view, or you may like to know about the architectural plan of the Taj Mahal; there are close-ups of engraved floral inlay work on the Taj Mahal, which one visiting would never get to see. Still, thousands of people travel thousands of miles from all over the world to visit the Taj Mahal. It is not about 'seeing' the Taj Mahal; it is about 'feeling' it. It is about immersing into the whole concept of the Taj Mahal by feeling it within its broader context. While appreciating its architectural marvel, a visitor to the Taj Mahal would like to do so within its contextual precincts. Expert advice is sought by a visitor to build this connection.

An increasing realisation is that tourism, today, is mostly about experience. It is about understanding a different context and an appreciation for the natural and built assets at a destination. Therefore, a single tourism product has many dimensions and holds a different meaning for different individuals. Moreover, different individuals may be interested in different aspects of a destination being visited. Nevertheless, it is abstract and conceptual at best. Typically, only a fraction of the tourism product is visible and tangible. A more significant part is ephemeral and experiential. It is, therefore, a challenge to sell a tourism experience. The challenge is compounded by the fact that resources often constrain visitors. They have limited time and financial resources. Further, being from a different cultural context, they have minimal access to the life of the host community, which is a rather important component of the context and is therefore essential for any interpretation to make sense. Visitors, therefore, seek expert advice as part of the tour package to make sense of the destination being visited.

One of the experts at the destination is often the *tour guide*. A guide's job is not just to tell visitors about a monument or destination; it includes helping visitors understand and feel the place. The tour guide is responsible for creating an image and a proper context for visitors, and they should facilitate learning for visitors. Therefore, a tour guide is often called an *Interpreter* or a *Tour Leader*.

Tour guide – the mediator

For a visitor, a place is a recreation of a narrative bound by his/her imagination. However, this narrative is built through multiple mediations. Cultural and ideological structures of diverse intermediaries shape the tourist's experiences. These include "government officials, tourism planners, advertising and marketing agencies, associated 'hospitality' industries ... travel agents and guides, travel writers and publishers, preservationists, and even people who study tourism" (Chambers, 2001). Tour guides are significant intermediaries who spend a significant amount of time with the visitors who lend a conscious ear to them. Ap & Wong (2001) suggest that tour guides have two vital roles – mediating and cultural broking as interpreters of the place. Tour guides mediate between tourists and locals and the environment. Mediating goes beyond merely telling visitors how to think and feel about their experiences; it is about helping them to arrive at their conclusions and learn in the process. The mediation role of guides has different meanings. He mediates between hosts and guests; mediates between the tour operator/ travel agency and the tourists; mediates between the tourist and the local community and local scene; mediates between the accommodation sector and the tourist; mediates between the local business including transport and tourist; and many others. Gurung, Simmons, & Devlin (1996) also hold a similar opinion that a tour guide builds bridges among different groups of people through the deployment of money, services, access, and information.

Gurung et al. (1996) also distinguish between tour management and experience management. They appear to agree with the concept of different roles for tour guides, as described by Pond (1993). They also suggest that a guide's role should involve tour and experience management – the organisation of itinerary and selection and interpretation of sites in an interactive process with tourists.

Thus, on the one hand, tour guides serve as cultural brokers whose pivotal role is to shape the visitors' impressions and attitudes and help enhance their appreciation and understanding of their surroundings. Cultural broking is the act of bridging, linking or mediating between groups or persons of differing cultural backgrounds to reduce conflict or produce change (Jezewski & Sotnik, 2001). For example, Ap and Wong (2001) say tour guides, through their knowledge and understanding of a destination's attractions and culture and using their communication skills, transform tourists' visits from tours into experiences.

On the other hand, they serve as a buffer, insulating many tourists from the difficulties (and possibly some delights) of the visited culture. Tour guides, thus, significantly shape a visitor's perception of a place. One of their essential roles is to encapsulate the spirit of the place and help visitors internalise its essence.

As per the conventional idea, guiding is considered a one-way communication, with the guide only telling about the destination. However, the scenario has evolved in the contemporary world, and it has become more interactive than in the past (Kainthola, Tiwari, & Chowdhary, 2021).

A brief history of tour guiding

Tour guiding is one of the fundamental activities in a tourism system where visitors negotiate a place with professional support. Many people believe that tour guiding is one of the oldest activities in the world (Cetin & Yarcan, 2017). The Greek historian Herodotus has documented the prevalence of guided tours in ancient Olympia in Greece and Giza in Egypt. (Dewar, 2000) cites Herodotus as referring to two specific roles of those guides – *peregetai,* which was orienting or leading the visitors, whereas *exegetai* referred to explaining. Weiler & Black (2015) also noted that a large number of statutes and plaques in ancient Greece necessitated professionals to explain their significance to the visitors. Then, there were specialists called *proxemos* whose job was to assist people in travelling abroad.

Descriptions and explanations that are available in ancient historical records cast an aspersion on their correctness. In fact (Batten, 1978) notes that these guides often exaggerated and included made up stories to ignite the imagination of visitors they accompanied to these sites. These stories do not match other available descriptions yet were considered fascinating and engrossing.

Ancient India also has a reference to guides assisting travellers. India was then a happening place. A common oft sited reason for travel was pilgrimage referred to as *tirthyatra*. People travelled to *tirthas*, the sacred places. There are references to people travelling from one *ashram* (hermitage) to another to meet *rishis* (sages) to learn, and even to have a peer review of their theories, or *granthas* (treatise) that they composed. These hermitages were often in remote places, away from urban or community settlements, like forests, mountains or on the banks of rivers. Then *Natya Mandalis* or the theatre groups roamed from one place to another to perform and entertain people. They often doubled up as guides and messengers to people who wanted to connect with or look for the missing people. Travel for leisure and pleasure has been little reported. Some guides lead these people through forests and unknown terrain. *Satvahak*, the caravan head, was an expert navigator leading the group (Goyal, 2020).

The fall of the Roman empire saw the advent of the Middle Ages or the Dark Ages. The contemporary world slipped into disarray. There were smaller kingdoms in Europe, and travel was adversely affected. The vast road network was in poor condition, and travel was unsafe. As a result, fewer people were willing to travel for leisure. However, this period saw the rise of Christianity in Europe, which prompted the rapid increase of pilgrims travelling to sacred Christian sites. Mysterious religious sites and biblical stories motivated devouts to travel and discover their Christian roots. These were the primary audience for the Middle Ages guides who would have delivered a historical snippet highlighting the sacredness of the place. The guides of the Middle Ages not only disseminated information but also ensured and guided safe passage. They paid bandits protection money and led pilgrims through mountains and forests.

Throughout the 17th to 19th centuries, elite European youngsters, especially the British nobility and landed gentry of sufficient rank and means, undertook a customary trip to Europe mainland, including France, Germany, Austria, Switzerland, and Italy, which were considered the seats of cultural resurgence. These 'grand tours' were taken by very young men accompanied by caretaker teachers or interpreters who were clergy or schoolmasters (Dewar, 2000). *Cicerone* was a personal guide tutor. Later young ladies typically accompanied by a chaperone also started undertaking the grand tour. During the three centuries of grand tours, there are several references to tutors who made a living by shepherding and educating young aristocrats on these educational trips.

Brito & Farrugia (2020) refer to guides who lived in what is now Italy in the 17th century. Called *Venturini*, they drove English boys and their tutors in four-wheel vehicles (*Vetture*) from one place to another. Besides transporting them, they also took care of their physical protection. Over the years, tour guiding has been there in different shapes and shades.

Towards the end of the 18th century, Europe saw the emergence of the Romantic era. Between 1800 to 1850, the advent of the artistic, literary, musical, and intellectual movement was partly a reaction to the Industrial Revolution.

Travellers in this time wanted to escape this industrial urbanisation for unspoilt serene places in the mountainous Alps of Europe. As a result, the first inn of this region opened in 1765 to cater to the growing number of nature-based tourists. New to the terrain, the visitors found first guides in the local hunters to lead them into the mountains. These local hunters were also knowledgeable about nature, told entertaining stories, and shared exciting information (Morgan-Proux & Cable, 2018). Soon these guides were offering one-hour and nine-hour day trips to the valley forests and along the streams. Thus they were the first nature tour guides who prioritised natural heritage over cultural heritage.

In 1821, the Compagnie des Guides de Chamonix was formed to regulate access to the mountain slopes co-operatively owned by the community. This guide association held a monopoly of guiding until the French government broke it in 1892. After this, guides were required to have a diploma issued by a commission that represented civil servants and the French Alpine Club members rather than residents. Today, nearly 200 years later, the Chamonix Guides' Company still exists.

Modern guiding and interpretation began to emerge in North America in the latter half of the 19th century. There is an interesting reference to a slave cave explorer, Stephan Bishop, who explored and showed the Mammoth Caves of Kentucky between 1838 and 1857. Almost the same time, in the 1830s, a good number of missionaries and traders were arriving in Rotura in New Zealand. These visitors were awed by the geothermal environment of the place, and local Maoris were quick to make use of this interest. They started serving as local guides to famous silica terraces in Rotomohana. Tourists were taken to these terraces in large canoes called *Te Arawa,* until the terraces were destroyed in the Tarawera eruption of 1886. There are references to two English speaking tour guides, Sophia alias Te Paea Hinerangi and Kate alias Keita Rangitūkia Middlemass, who used to get 15 shillings from each tour party (NZ History, n.d.). In the 1880s, tourist interest shifted from Tarawera to Rotorua. A toll was of two shillings and sixpence was charged from visitors to the village, which included one shilling as the fee for the guide.

At the turn of the 19th century, Americans grew conscious of preserving nature and enjoying it. Subsequently, the Yellowstone National Park was opened up in 1872. Guests travelled by train to the park and usually stayed longer for several weeks. A visitor could be lost in the vast expanse of wilderness, rough trails, and dangerous thermal features. Thus, the first tour guides in Yellowstone worked as stagecoach drivers and thermal-basin walking guides

The Battle of Gettysburg concluded in 1863 with the most significant casualties in the American Civil War. Soon after the war, grieving family members and curious visitors arrived on the battlefield. The residents took advantage of this emerging situation and began to escort guests to the different sites of the battle. As more visitors started visiting the park, the demand for tour guides increased, and by 1915, more than 100 tour guides were trained for battlefield tours. Even so, tourist guiding was still not a profession. Guides charged different fees and narrated stories that were not necessarily true. To improve the quality of the guides, the government began a process to license official guides for the battlefield tours and educate them about the regulations covering fees, conduct, length of tours, and grooming. Thus, the Gettysburg Licensed Guides became the first professional guide service in the United States.

In 1920, Enos A Mills published *The Adventures of a Nature Guide* which introduced the profession of nature guiding and offered the basic definitions and philosophies which became the foundation of the National Park Service interpretive program. Later in 1957, National Park Service engaged Freeman Tilden to improvise training of professional nature guides based on these principles. Freeman Tilden (b. 1884) is often credited with contemporary guiding. His book *Interpreting Our Heritage,* published in 1957, promulgated the *six principles of interpretation,* and is still considered a seminal work.

Thus the ancient vocation of guiding has evolved into the modern avatar of professional interpretation. All through this time, guides have led visitors in search of meaning in material, intellectual, and spiritual sense.

Tour leader – guide, escort, or manager?

Often the terms tour guide, tour leader, tour escort and tour manager are used synonymously. Even the industry uses these terms vaguely sometimes to mean the same set of responsibilities while some other times a hierarchy among the roles is created. Let us demystify the role and job of a tour guide vis-à-vis the other roles.

In practice, a tour guide is a person who explains the site to a visitor. They usually deliver a commentary to the visitors at a site. They are expected to be experts in their area, city or region. Depending upon the situation, the role of a tour guide can be envisioned variously. The primary role is that of a site guide where they are experts on a particular site or activity. If their job is to interpret the only specific structure, they are called *monument guides* like a guide at the Taj Mahal or a guide at the Leaning Tower of Pisa. If it is an archaeological or natural site (like caves or some other geological formation) or a spot with a reasonable spread, they may be referred to as *on-site guides* or simply *site guides*. In the case of specific locales like parks, an adjectival prefix may be used, for example, the *park guides*.

Similarly, in the case of a museum, they may be called museum guides. Museums also use volunteers or docents to assist visitors in interpreting the displayed art and artefacts. Then there are *nature guides* and *adventure guides*.

These assignments range from a fraction of an hour to a few hours. This category of guides is the ones who get the minimum wages. *Museum volunteers* and *docents* sometimes work gratis. It is only in some cases that the site authority employs the tour guides on their muster rolls. For example, park authorities or archaeological site authorities may employ salaried guides. Some museums and private monuments also have salaried tour guides. Corporations and some big establishments may have guides on regular salaries for factory tours or, say, the tour of the parliament building complex.

Some tour guides expand their jobs to include multiple attractions in the city. They accompany guests for the day trip where they claim to be the experts of the village, the town or the city. Besides the places of interest, they interpret what is happening around the place. Hence, they are referred to as *local guides* or *city guides*. However, some people make a distinction between the two, and suggest that these city guides generally do not interpret the details of a monument. Instead, they point out the exciting places and share brief details about them while on a coach or tram tour through the city. Thus, they focus on the interpretation of the times and life of the city (or area) or a particular aspect of it rather than one monument or attraction. Variants of this include bicycle tours, rickshaw tours, and others. More recently, it can be a Segway tour. Another option that is becoming popular is a walking tour, and in these instances, the guid is known as a *walk leader*. The other guides, who interpret many monuments and sites within a day's duration are *local guides*. A limiting case of a local guide can be a single location guide or the site guide.

Generally, this guide's job ranges from a few hours to a day's length – up to, say, 12 hours. Their primary job is to be skilled in disseminating information about the locale interpreting it for the visitors. They lead the group at the place and are therefore sometimes known as tour leaders. In some organisations/ countries, the tour leader's role includes overseeing travel arrangements. Moreover, this is so engaging that they hire a professional guide to interpret some of the places on the itinerary. Thus, they tend to become tour managers. However, a tour leader is often two-in-one, acting as tour manager and guide on the same tour or different tours.

While a tour guide or tour leader narrates the tours and leads the group, a company representative often accompanies the group to coordinate that every little thing runs smoothly. He/she is referred to as a tour escort. Their job is to attend to logistical details. In some companies, this role is designated as a tour manager. However, there is a subtle difference of authority between the tour escort and the tour manager. Often the escort attends to tactical jobs and is subordinate in authority to a tour manager while they might appear to be performing the same job as a tour manager. Tour manager, however, is a more responsible position. Sometimes, the tour guide and escort report to the tour manager. The term 'escort services' has become notoriously prevalent for depraved activities. This term 'escort' is, therefore, falling out of usage. Therefore, the industry is increasingly using the term tour leader, tour supervisor, or tour manager.

A tour manager is responsible for the conduct of the tour over its entire duration. Thus, the responsibility of a tour manager spans beyond a day's extent. He or she is hired to be with the group throughout the trip from arrival till the party's departure, and to handle all the aspects of the tour.

Further, different companies organise themselves differently. In many cases, there is an overlap between the duties of a tour guide and a tour manager. In some cases, the tour guide is on board at different sites or for a day. However, in other companies, the tour guide is with the group for the entire trip duration, ensuring that the tour manager's itinerary and activities are completed as conceived and planned. In other instances, the tour manager accompanies the group throughout the travel but do not deliver any commentary. Instead, their job is to attend to the details and take care of the clients.

In other companies, the tour manager's job is back-office. They operate from the office. They conceive the tour and work out the finer details. They finalise the dates, the routes, connecting flights, trains, buses, stay, transits, sightseeing, and all other details. They have the overall responsibility for the tour, ensuring that everything falls in place and there is a seamless tour even if they are not with the group on the trip. Even if the tour managers are on the tour, they generally do not narrate it. Their job is only to extend logistical support. In some markets, terms like *tour conductor* and *tour-courier* are also used.

In India, there is a three-tier tour guiding system. At the lowest rung are the monument guides or the site guides whose scope of operations is restricted to a select site or monument. Site guides, local guides and city guides constitute the first level of guides in India.

Some guides, over time, become familiar with the places of tourist interest in more than one city. In India, the second level of tour guides are the state-level guides. It is assumed that they are experts of the tourist attractions dispersed across the state (a province). The state-level DMOs licence them. The itinerary of the visitors extends well beyond the city and the day. Thus, the guide escorts the group through several days. The companies, therefore, also expect them to take good care of their clients. The job expands well beyond that of narration. At the top of the ladder are the Regional Level Guides who are licensed and therefore can accompany the clients across several states in the region. Realistically, they cannot be experts in all the places in the area. Over a while, these guides acquire expertise about the principal monuments in the area. However, their job gradually evolves into carer of guests as the face of the company on the tour. As and when required, they hire the services of the resident site or local guides.

A tour guide (US) or a tourist guide (European) is a person who guides visitors in the language of their choice and interprets the cultural and natural heritage of an area. This person normally possesses an area-specific qualification usually issued and/or recognised by the appropriate authority (Definition as per EN 13809 of the European Committee for Standardisation adopted by WFTGA, CEN, 2003).

Typically, in India, a beginner starts as a tour escort for some tour operators and acquires valuable experiences and skills. To accompany the guest to all the monuments, they must obtain the necessary tour guide license. Then, they must gain the necessary qualifications for the same. After that, they may be licensed as a state-level or a regional tour guide. The rules in India require that only a licensed tour guide may accompany a tour party into Archaeological Survey of India (ASI) restricted monuments. Some other escorts prefer to hone their tour management skill to be placed as professional tour managers.

Thus, while the jobs of a tour guide and a tour manager are conceptually different, their roles often overlap. For this book, we will use the terms ***tour guide*** or t***our leader*** for a professional responsible for commentary and interpretation with a limited managerial role. However, we will differentiate between the two roles when the need arises.

What do guests want the tour guide to do?

Erik Cohen (1985) was among the early researchers to have discussed the role of a tour guide. He opined that tour guides should display high-level professional skills and intimate knowledge of the local culture. According to him, it is the job of a tour guide to "translate the strangeness of a foreign culture into a cultural idiom familiar to the visitors." Thus, they have two crucial roles to play. One is the pathfinder role, where the guide provides access to visitors to otherwise non-public territory. The other role is a mentor or 'guru' who offers "insight, enlightenment or any other exalted spiritual state".

Cohen (1985) suggested four types of guides who primarily carry out four different functions – instrumental, social, interactionary, and communicative. The first type is the ***Originals***. They are the pathfinders who focus on the 'instrumental' function and ensure that the visitors get to the desired destination and return safely. Some refer to them as *Path Breakers* as they chalk out the route linking the various attractions at a destination and ensure access to them. They focus on completing the itinerary while offering limited information about the attractions. They keep away from the detailed interpretation of the places and things. The second type of guide is the ***Animator***. They typically attend to the social dimension of the interaction by intermingling and socialising with tourists, being friendly, listening and respecting their preferences. They try to make visitors comfortable and feel relaxed. The third type proposed by Cohen is the ***Tour guide***. They perform the interactionary function by enabling interaction among tourists with the locale. In addition, the tour guides must mediate. The last types are ***Professionals*** who are like mentors. Tour guides as professionals perform a communicative function. This communicative function of the tour guide has four components: (1) designing the itinerary (suggesting what to see and experience – also what not to see); (2) dissemination of information that is correct and detailed; (3) interpreting the experiences; and (4) fabrication of information, that is, presenting false information as if it were genuine/real.

Reisinger & Steiner (2006) suggest that tour guides have two essential roles: transferring detailed information (e.g. telling to tourists where, when and why to look, how to behave); and interpreting attractions, sites and experiences.

A tour guide is the company's face for the guests on tour. The guests expect the tour guide to be the single-window solution to all their concerns.

1 ***Guest are here for relaxation***. While they relax, they want the tour guide to care for their worries. So a tour guide must ensure that the guests are kept away from all the botherations. Even if there are challenges that come up, and they will come up, keep guests away from them.

2 ***Guests need not worry about logistical details***. The tour guide and the ground crew must ensure that all things are in place – the tickets, the coach, the transfers, the vouchers, the site guide, any and everything required. A tourist might be visiting the place for the first time on a limited duration holiday. They would not like to waste their time and expend their energies on any of these petty things.

3 ***Achieve the itinerary***. A guest has purchased sightseeing to all the places listed on the brochure. The one thing that will offend the customer the most is missing an attraction listed on the brochure. A tour guide and a professional ground crew must ensure that the clients visit every listed place and are still left with free time to shop and explore the place independently.

4 ***Guests expect an enjoyable holiday***. They want to enjoy their travel and social interactions with others in the group. That is often the reason for going as part of a group. The tour guide is the link pin and must promote acquaintance and interaction among members. They may engage willing guests in activities and programmes – but not force them.

5 ***Source of information.*** The tour guide is the ready reference for the clients. They expect him/her to be knowledgeable about the places, things, details, procedures, local places, people, local tales, and everything that is not in guide books or the public domain. Therefore, tour guides must have visited as many places as possible earlier and have first-hand information about these locations.

6 ***Expect respect.*** Guests want to be treated with the utmost respect by the tour guides. Guest come from various backgrounds, and many of them are first-timers visiting a foreign land or on a conducted tour and may not be aware of the nuances or protocols. Therefore, the tour guide needs to empathise with them and remain courteous at all times.

What do companies expect of a tour guide?

Tour guides are usually independent professionals who work with several different companies simultaneously. Though the tour guides are not usually salaried employees, they are the company's face for the clients. For a guest, they are the company. The company, therefore, relies to a large extent on the tour guide for the success of the tours. Tour guides and the ground crew are responsible for delivering a superior experience for the clients. The ground crew needs to fructify all the hype created by sales and marketing. A successful tour generates positive word-of-mouth about the company, which is the most reliable promotion. The companies, therefore, expect tour leaders and tour guides to conduct themselves as if they are the company:

1 ***It is all about an entertaining and enjoyable trip.*** Guests buy a once-in-a-lifetime holiday. They want the company to take care of the logistics and other details while they have uninterrupted fun. The job of the tour guide is to ensure this. Guests and their fun must be the priority of a tour guide.

2 ***The tour guide must always sound positive.*** They must be friendly and considerate towards the fellow crew and other partners (vehicle, staff at attractions, and others) and empathetic towards guests.

3 ***Tour guides must conduct themselves professionally.*** All assigned duties must be carried out meticulously and with responsibility. Their behaviour and grooming must be of the highest standards, refraining from smoking and consuming alcohol.

4 ***There will be unprecedented situations.*** Therefore, a tour guide must always maintain calm.

5 ***A tour guide, to succeed, should have the support of the entire team.*** They must always be ready to seek and support the operations team, sales team and ground crew. Tour guides must display loyalty towards the company.

The job description of a tour guide

A job description summarises a given role's primary duties and responsibilities in a company. For example, based on the expectations of the tour operator and their guests, the job description of a tour guide typically includes the following:

1. Organise the trip as per the details and schedule.
2. Ensure that all sites listed in the promised itinerary are visited.
3. Brief the guests in detail about the daily programme and share the outlines of the program in the coming days.
4. See that the guests are safe and secure at all times and raise their awareness of safety and security issues.
5. Validate that the coach is fit for use and advise guests about the safety protocol to be observed on the bus or other moving vehicles.
6. Share details of facilities and arrangements for their comfort and convenience, along with basic do's and don'ts on the tour.
7. Narrate details of the important landmarks as the coach passes by. The tour guide must deliver comprehensive descriptions during the excursions.
8. Engage with the local guide as and where required (as per the arrangements made).
9. Confirm that meals at a restaurant (or elsewhere) are organised as per promised provisions and that the meal requirement of each guest is taken care of.
10. Regularly communicate with the ground crew and coach pilot to plan for the next moves and keep them updated about what has been achieved.
11. Maintain proper accounts, including item-wise details of all expenses. Safely retain receipts of all expenses and ensure submission and settling of the accounts on completion of the tour.
12. Keep an eye on expenses and try to minimise them without compromising the quality of the experience for the guests.
13. Not promise additional incentives or refunds without authorisation from the operations department.
14. Guarantee that the tour is conducted responsibly. At all times, the tour guide will ensure that the group is sensitive to the surroundings, including the transport and accommodations.

Job specifications for a tour guide's job

A job specification details the abilities and qualifications required of a person for the role. Based on the job description, a tour guide is expected to be able to perform the following five functions:

Manager

A tour guide is primarily a manager of things and situations. They must plan and execute hassle-free daily programmes for the group, mobilising all necessary resources. They must keep the guests informed. They must make sure that lodging arrangements are in place, carry the rooming lists and see that meals are as per the guests' preferences. They must make sure that proper passports and visas are there for frontier formalities, and carry all necessary documents – passenger lists, vouchers, contracts, insurance policies, route maps, and any other. They must complete all documentation related to visits and expenses.

Performer

A tour guide is also a performer whose job is to keep the guests entertained. Their primary job is to provide helpful information and exciting commentaries throughout the tour. In addition, they need to build a rapport with the guests and keep them engaged by facilitating unique experiences and organising events and activities for them.

Attendant

A tour guide is a caretaker of things, situations and individuals. Most guests expect tour guides to ensure that they enjoy the trip without any challenges. The guides need to ensure guests' safety throughout the tour. They must advise guests on how to keep themselves safe and secure. They must handle emergencies arising from illness, thefts, accidents, loss of belongings, and other unforeseen situations. Furthermore, they must get quick, professional assistance – police, diplomatic, medical, or whatever is necessary for such situations.

Tactful

A touring party is often a group of people who might not know each other. Each person has their quirks. Keeping them together and accommodating their peculiarities requires a tour guide to be tactful. They must demonstrate tact in handling untoward situations. They should also manage relations with the guests, the company, and the various suppliers.

Accountant

It is the responsibility of a tour guide to handle on-tour expenses. They may be conservative with on-tour costs. They must pay all bills and tolls when required (and authorised). They must keep enough cash for exchange rates and charges. They must maintain accounts meticulously and submit them on completion of the tour.

Technology and tour guides

The tour guides are often sceptical of technology which they believe would displace them. Conventionally, they maintained that included in the role of a tour guide is disseminating information related to the site and destination. However, tour guides realised that guests are empowered by technology and have instant access to information. Therefore, the tour guides' role is shifting to facilitating the visitors as they immerse in experiences at the destination.

Covid restrictions have severely impacted tour guides, who often worked as freelancers and relied on gratuities and commissions. With tourism coming to a standstill, their earnings have taken a beating. Cautious guests, on their part, are avoiding contact with other people, including tour guides, as they travel again. In addition, travel agencies and tour operators are redesigning their offers with minimal human contact. Instead, they are substituting human contact with technology. Tour guides consider this a threat.

The 'untact'

The need for social distancing in Korea led to the emergence of 'untact', which meant 'no contact.' This concept also influenced *no contact tourism, which refers to tourism that avoids crowded places* or indoor activities and encourages visiting outdoors that are sparsely populated.

It is expected that the Covid pandemic will hasten the use of technology, especially digital applications. Therefore, the Korean government shifted to technologically leveraged tourism, focusing on 'smart,' 'personalised', and 'non-contact travel. Many tourists were hesitant to travel accompanied by a tour guide, so the Korean Tourism Authority developed the 'Digital Storytelling Service', substituting in-person tour guides. Visitors can download the 'smart tour guide' application on their mobile phones. This app allows interaction in languages of their major markets – Korean, English, Japanese, and Chinese. More languages are to be added subsequently.

Furthermore, the Korean Tourism Organisation has launched a website Durunubi (www.durunubi.kr), that provides information about trails and treks nationwide, which will be helpful to the growing number of tourists seeking outdoor activities. This service also provides information on the routes, including popular tourist attractions, cultural heritage sites, transportation, accommodation, and restaurants (OCED, 2020). Thus many responsibilities that tour guides used to own are now offered technologically.

The pressures of Covid-19 catalysed the use of technology. As a result, tour guides have begun to embrace technology to promote their businesses and enrich guests' experiences. (Brito & Carvalho, 2021) point out that post-Covid-19, tour guides have gotten innovative. They are now using new technologies to promote themselves as entrepreneurs. They are also using audio devices and electronic props on their tours. Weiler & Black (2014) noted that the tourism industry will increasingly use technology in marketing and delivering experiences.

Benefits and limitations of using technology

There are pros and cons to using the digital applications have been developed to assist tourists (Reed, 2015). One advantage that stands out is that apps present well-researched and factually correct information. Today, visitors do not need volumes of factoids. Instead, as per their interest, they search for them online. Moreover, this information is available at all times. Generally, the information is cautiously curated so that the content is rich enough to help visitors develop a perspective on the relevance of the place and the history surrounding it. This overview goes a long way in aiding self-drive travellers to experience what the place has to offer. Moreover, the tourists can visit the place at their own pace and, as per their route, skip sites and attractions that they are not interested in, thus customising their experience. This flexibility is otherwise constrained if accompanied by a tour guide on a somewhat fixed itinerary.

Another advantage of the digital app is its ability to mass customise. Further, many consumers can log into the app simultaneously, while the interface can be available in many languages. Further, the apps usually can share consumers' experiences or comments on the services or the attraction on their social network. Similarly, the guests also have real-time access to comments and opinions shared by other users, thereby giving them a richer perspective on what the place or attraction offers. Further, the cost of guiding the visitors is less for the tour operators who substitute tour guides with these apps.

It is generally believed that the tour guide offers a richer experience, customising interpretations, handling spot requests, and making last-minute changes to match the guests' expectations closely. In addition, most tour guides possess a personality and charisma, which is pivotal to the visitor's experience, while the technology lacks personal interaction with the local community, which tour guides offer. Interaction with a tour guide is also an opportunity to engage with a local who adds to the experience by bringing an indigenous perspective to the interpretation. Many visitors believe that personal interaction with local tour guides is far better and far more interactive as they enthusiastically receive queries and questions which they respond to. As visitors receive the information, they usually have follow-up questions that the apps would not answer. Thus, tour guides offer a personal touch to the interaction with the guests. While the apps provide a range of validated information, tour guides make up by offering interesting additional anecdotal information.

They also mediate between the destination and local community and the visitors, wherein their narrations can add meaning to a tour. In addition, while the apps cannot go beyond their defined jurisdiction, the tour guides can go out of the way to assist in transfers, health concerns, missing documents, seeking advice for shopping, etcetera. Also, the tour guide, being a local, has an extensive network with local service providers and can help accommodate requests – however unique they may be.

Technology does offer efficiency, but using apps is like being served by vending machines. Apps and their protocols may make their visits mechanical and mundane. Visitors often prefer to explore the persona of the place through the guidance of a professional who is trained and qualified to provide the relevant and curated information and is charismatic enough to deliver an unforgettable experience. Reporting in *Travel Weekly* (Baran, 2014) quotes service providers that visitors want to learn more about lifestyle characteristics of a place; for example, besides visiting the Taj Mahal in Agra or Coloseum in Rome, they are also keen on tasting local cuisine. She quotes Paul Niskanen, a Travel Advisor:

> "In the past, guides focused more on facts and figures. Today, a guide becomes more of an 'instant friend in Italy or Spain,' in that he or she talks a little more about his or her family life, good restaurants and local traditions."

While the costs of guiding guests could be less for the companies, many responsible visitors engage tour guides because it has a direct positive economic impact on the local community from which the tour guides are usually drawn.

Apps, in all certainty, will replace some components of routine commentary and information dissemination. Therefore, technology will be increasingly relied upon despite the richness that tour guides bring to the experience. It is time for tour guides to move on and embrace technology and acquire the necessary skills to use this technology for delivering superior experiences. For their part, visitors will use technology and apps combined with live tour guide commentary to contextualise what is being presented. However, the proportion in which the travellers will use apps and engage tour guides may vary. Life skills, social skills, and guiding abilities acquired with much effort will still be valued. (Reed, 2015) is of the view that apps may not wholly replace tour guides, who personalise the interaction and offer a unique perspective to the place.

A good tour guide is good if they serve guests to their satisfaction, irrespective of whether they use technology or not and how they use technology. So the superordinate goal is to provide excellent customer service each time, every time, and whenever. Technology might come in handy to serve a guest, but they would still appreciate non-tech-savvy tour guides for their personalised services.

The tech-savvy tour guide

Today, a tour guide travels with a smartphone and tab with high-speed internet connectivity. They use technology for efficient service and compelling experiences. In addition, the devices are loaded with apps that increase their productivity – reporting systems and the company's learning management system and proprietary online training, and instant connection through messaging apps to keep the guests well informed of up-to-the-minute itinerary development.

Their devices, to assist guests, may have apps that include mobile payment and credit card processing, Google Maps and GPS, weather forecasts, currency conversion, etc.. In addition, they regularly access the Web to book restaurants and shows for guests, and to check train, bus, ferry, and flight schedules.

For augmented guest experience, the devices may be loaded with various resources ranging from information about the destination, including images and presentations. They may also carry music, videos, animations, site plans, drawings and explanations. A creative tour guide innovates with their gadgets as much or as little as they want to curate unique experiences for their guests. For example, some use these gadgets and apps to customise welcome letters, handouts and advice. Some even use apps to create photo slide shows to present to guests at the end of the tour.

Another area where technology has made a mark is in the realm of on-ground communication. Tour guides now use wireless headsets to connect with their group. This technology has changed how tour guides today can remain connected as they walk, talk and interpret the site. While visitors can roam and take photos at their own pace, the guides continue with their commentary without necessitating the visitors to hoard up around them. Therefore, they no longer yell or jostle for sound space with other visitor groups in the vicinity. Instead, they keep a quiet watch on their guests from a distance, speak coolly into their microphones, and have their entire group hear them.

Compensating tour guides

Tour guides generally work as freelancers, independent contractors or temporary workers who may or may not be associated with travel companies. In limited cases, they are on the rolls of the companies who pay them regular salaries. Freelancing tour guides get fees determined in different ways. In some cases, market forces determine the fees based on the individual's skills and experience. In other cases, the guide's fee is determined by trade associations. For example, in India, the Indian Association of Tour Operators (IATO) and Tourists Guide Federation of India (TGFI) periodically enter into an agreement to determine the minimum fees for the tour guiding assignments considering variables like the size of the group, language allowance, outstation excursion allowance, conveyance allowance, food allowance and extra allowances (IATO, 2019). In many other parts of the world, tour guides' compensation is tied to the daily wage rate, depending on the local cost of living. Thus, the wage rate is usually higher in bigger cities than in smaller places. For example, in New York, a tour guide can expect to get $40-50 per hour, whereas the same guide may get around $20 in a smaller market. Further, variables like experience, performance, training and qualifications as a professional tour guide also impact the compensation. The compensation of a tour guide also depends on the efforts invested in networking, acquiring skill and commitment one has to pursue this career.

Besides professional guides, many casual workers are often engaged in tourist guiding. Some are students who put in a few hours for some extra money for sustenance. Others are retirees who volunteer out of their love for history or the place. Then some workers engaged elsewhere, like waiters, bus conductors, or actors, are engaged as part-time guides by tour companies for evenings or weekends beyond their regular jobs. These are generally paid less than the professionally trained full-time tour guides. For them, it is a supplemental income.

In an open market, the compensation is also determined by the supply of guides. If many tour guides are available, they compete for guiding assignments undercutting their compensation and making it up by a higher turnover, that is, handling more groups instead. Consequently, travel companies in some markets tend to infuse under-trained or part-time workers to increase the supply and reduce the wage rates.

There are restrictions on engaging unlicensed individuals for the job or even an assignment at some destinations. In mature markets, licensed or certified tour guides are available at higher prices. Similarly, the demand-supply equation may differ in some sub-markets, especially in language-determined markets. For example, at some destinations, guides who can provide commentary in a particular foreign language are in short supply and are thus available at higher prices. In India, most trained guides speak English, while European linguist guides (Spanish, Italian, French or German) are available at higher prices. On the other hand, tour guides who can speak east-Asian languages like Chinese, Japanese and Korean are in short supply and are often available at three to five times the price of a regular tour guide. Similarly, specialist guides, like those who can accompany the guests with special needs or those who specialise in a particular product like food tours or art tours, come at an additional cost.

However, fees are not the only revenue stream for a tour guide. The tourism industry is competitive, and operators try to keep the costs less. Therefore, the tour operators are often conservative in salaries, especially for outsourced staff, including the tour guides. Tour guides, to make up for these meagre salaries, find other ways to supplement their incomes, and earn significantly from gratuities and commissions. However, gratuities and commissions work differently in different places and are culturally determined to a large extent. For example, tipping a tour guide is almost a norm in some cultures. Therefore, the tips are factored into tour guide fees by their employers.

In some cases, the group tour manager advises them and collects nominal tips for guides who serve the group at attractions and sites. As a limiting case, there is a 'free tour' concept. Here, there are no prices for the tour. Instead, tours are experiences that are veiled-priced as a pay-what-you-want model where tourists who take tours tip confident tour guides whatever they wish at the end of the tour. As this model evolves, guides increasingly pay a nominal fee to the tour operator to assign a group. They expect to make money by collecting tips from this group. On the other hand, there are no expectations for tips in some countries, and therefore, a higher wage will be paid entirely by the tour operating company.

Another important revenue stream for tourguides is 'commissions' on the purchases by the guests. Tourists often spend a lot on shopping and expenditures in restaurants and other activities at the destination. These guests, in turn, are led by tour guides who are better situated to advise them on local specialities to be tasted or taken back. Merchandisers and service providers, therefore, intend to rope in this very formidable 'gateway' as channel partners for sales. They incentivise tour guides with attractive commissions for recommendations. It is believed that the system of incentives started in Middle Eastern countries like Egypt and Turkey and subsequently spread worldwide to almost every tourist destination. In some cases, commissions and kickbacks are the dominant revenue streams for tour guides.

Cuts and commissions are not just a third world phenomenon. Escorting high net worth guests to luxury shops are handsomely rewarded in first-world markets also (Gebicki, 2018). Conventionally, a 10-35% cut is provided for the network partners as an agreed practice (Murali, 2017). Tour guides, coach and taxi drivers get their share of commissions as prevalent in the market. In some cases, tour guides get a flat fee, in addition to commission for bringing clients to stores or workshops.

Commission regimes, being a lucrative revenue stream, often shift the focus and interest of the tour guide from site interpretation to pushing for sales. This jostling adversely affects a visitor's experience harming the cause of tourism at a destination in the longer run. Many countries, therefore, ban tour guides from making purchase recommendations and accepting commissions. However, this is perfectly legal in other countries. Thus, despite bans and restrictions on commissions, this functions in the tourism ecosystem and serves as a revenue stream for tour guides.

However, there is an alternate view. While advising on local shopping, a tour guide also serves as a local consultant. They may get guests a good bargain, a genuine product or an authentic experience. Tour guides work extra hours beyond those agreed upon to extend valuable advice and get the visitors to the right place, sometimes at odd hours. This extra effort beyond the call of duty is not always equitably compensated by tips, and so a modest commission is arguably in order if the guests get what they want and the merchandiser or service providers get the right customers. A saving in the marketing costs of the vendors is partially passed on to the customers. This arrangement is a typical win-win situation for all the parties involved – the guest, the seller and the tour guide. These reasonable commissions that unduly do not inflate the selling price of the product indicate a healthy reciprocal relationship between the three parties, which in the end benefits the guest, any way you look at it.

However, of late, tour guides have been lamenting that while they bear the brunt of anti-commission drives of the authorities, they often do not escort guests to shopping establishments of their own volition. Instead, the tour operators who hire them direct them to lead guests to select outlets only. Thus, tour guides have been expunged from this system of commissions and, tour operators are directly involved.

Cross-selling has also emerged as a potential revenue stream for tour guides. Many companies that offer budget tour packages offer their guests add-on activities at an additional cost. They encourage and expect the tour guides and leaders to cross-sell these activities. Companies pay commission on the sale of these add-ons. However, some tour operators consider cross-selling a routine component of a tour guide's job description, and no additional incentive is offered. As mentioned earlier, tour operators and travel agents share a significant portion of the kickbacks from the sale of these products, and a small amount of that income is divided between tour guides and transporters who execute that optional activity on the ground.

This system of incentivised selling is prevalent around the world. However, only a few countries have regularised this within the ambit of a legal framework. Whereas in countries like Turkey, Egypt, China, and India, the tour operators are contracting with the craft places, restaurants, and attractions for additional incomes to sell their tours at more competitive prices. For example, in Turkey and India, some tour operators sell tours heavily subsidised by attractions and craft places.

Besides revenue streams, other benefits come a tour guide's way. They network with several service providers to informally help them. They return favours. One common incentive is paid for board and meals on long tours and complimentary lunch or dinner on day tours. Even when these tour guides are on private tours or get their families and friends for dinner at restaurants, they get hefty discounts. Similarly, merchandisers also oblige tour guides who have been helping them with customers.

In the case of tips and commissions, no employment relationship exists between the payer and the payee, the tour guide, and thus, this income is outside the purview of taxation in many countries. Also, because this revenue changes hands informally, this is not on record of the tax net. However, in other countries where commissions are legal, the transactions are on record.

Tourism is a crucial phenomenon today. Besides its economic contribution, it has socio-psychological implications. It requires professionals, including tour guides, to conceive, create and facilitate experiences for the visitors. Incidences like Covid 19 impact the patterns of tourism supply and demands, challenging the tour guide to relook at their vocation and evolve in a new avatar.

Key points

Tourism is an important economic activity, but post-Covid-19, it has reinforced itself as a socially desirable phenomenon. It is not about mere sightseeing; it is more about immersion in a destination's experience. Tourism today has become a way of life for many. It is travel for recreational, leisure, or business purposes, usually of limited duration.

It is not that people have started travelling now. On the contrary, they have been travelling all along for various reasons. Travelling is a basic human need. Superior infrastructure and technology are among the reasons why tourism has proliferated over the past few decades. An increasing realisation is that tourism is about understanding a different context and appreciating the natural and built assets at a destination. Therefore, a single tourism product has many dimensions and holds a different meaning for different individuals. Thus, it requires mediation by an expert to make sense of the place being visited. This expert is often the ***tour guide***.

Tour guiding is an ancient vocation. It has existed in different forms in different parts of the world since recorded history. Like many other professions, it has evolved from an ancient vocation into its modern avatar of professional interpretation. All through this time, guides have led visitors in search of meaning in material, intellectual, and spiritual sense.

Tour guides significantly shape a visitor's perception of a place; one of their essential roles is to encapsulate the spirit of the place and help visitors internalise its essence. Tour guides have two vital roles – mediating and cultural broking as interpreters of the place. Tour guides mediate between tourists and locals and the environment. Mediating goes beyond merely telling visitors how to think and feel about their experiences; it is about helping them to arrive at their conclusions and learn in the process.

Often the terms tour guide, tour leader, tour escort and tour manager are used synonymously. However, even the industry uses these terms vaguely, sometimes to mean the same set of responsibilities while a hierarchy among the roles is created. We have discussed the nuanced difference between these roles. We will use the term 'tour guides' for this book, including tour leaders (distinct from tour managers).

The chapter also discusses the guest's expectations from their tour guide. Tour guides translate the strangeness of a foreign culture into what the visitors understand. Thus, they have two crucial roles to play. One is the pathfinder role, where the guide provides access to visitors to otherwise non-public territory. The other role is a mentor or 'guru' who offers insight. Guests expect tour guides to be a single-window solution to all their concerns, while tour companies expect guides to represent them and handle guest concerns on their behalf. As a result, they rely on the tour guides to a large extent for the tour's success.

Based on the expectations of the tour operator and their guests, companies create job descriptions for the tour guide. We discussed some of these expectations in the chapter. Based on the job description, a tour guide is expected to perform five functions as part of the job specification – manager, performer, attendant, tactful, and handling accounts of the assigned trip.

Tour guides are often sceptical of the technology and fear that it might displace them from their jobs. However, technology is here to stay. Consumers will increasingly rely on technology for many aspects of their tours. The only way out is for tour guides to embrace technology and use it to their professional and personal advantage. However, they must also realise that acquired life skills, social skills, and guiding abilities will still be valued.

Finally, we discussed the issue of compensation for tour guides for their services. Usually, there are multiple revenue streams available to the tour guides – besides the guiding fees, guides benefit from gratuities and commissions on cross-selling and contracted sales. Unfortunately, the latter often leads to unethical practices in not-so-mature markets and is thus prohibited. Guides must, however, factor in these different revenue streams in their compensation plan.

Key terms

Commission	Kickbacks tour guides or travel operators receive from merchandisers and service providers who sell goods to guests brought to their shopping establishments, restaurants, accommodations or shows.
Fees	The agreed-upon payment made to the tour guides by travel agents or directly by the customers for their guiding services.
Gratuity	Tips paid voluntarily by the happy and satisfied customer to the tour guides in appreciation of good service. This perk is over and above the agreed-upon fees.
Mediation	Creating a rapport between the tourists and the locale by shaping their perception of the place, telling visitors how to think about their experiences and helping them arrive at their conclusions and learn in the process.
Tour escort	A company representative often accompanies the group to coordinate and ensure that things run smoothly. Their job is to attend to logistical details.
Tour guide	A person who interprets the site to a visitor, usually delivering a commentary.
Tour leader	Leads the group at the place. Their primary job is to be skilled in disseminating information about the locale interpreting it for the visitors, but in some cases, their role includes attending to scheduling and logistical details.
Tour manager	Person responsible for the conduct of the tour over its duration and hired to be with the group throughout the trip from arrival till the party's departure.

Discussion questions

1 Tourism has evolved beyond mechanical sightseeing. What do you think is the role of tour guides in creating and facilitating experiences for the guests?

2 Tour guiding is an ancient profession. It has evolved over the period. Please evaluate how the role of a tour guide has changed during this period.

3 Identify some traits you think are essential for becoming a tour guide.

4 Talk to some tour guides to understand how technology has impacted their job description. How do they cope with technological advances in their field?

5 Observe and talk to a few tour leaders to understand how they use mobile phones to their advantage in performing the assigned role.

6 Discuss with some tour guides operating at a popular site to understand the relative share of different revenue streams in their earnings? Also, understand their position and ethical dilemma vis-à-vis the expectations of the employing tour operator.

Activity

1 Form a team of three where one person creates the itinerary, the other person represents the tour guide, and the last person acts as a tourist. Execute the itinerary under class activity keeping in mind the different roles of a tour operator, tour guide, and a tourist's expectations.

References

Ap, J., & Wong, K. K. (2001). Case study on tour guiding: professionalism, issues and problems. *Tourism Management, 22*, 551-563.

Baran, M. (2014). *Technology presents challenges and opportunity for tour guides*. Retrieved October 23, 2021, from Travel Weekly: https://www.travelweekly.com/Travel-News/Tour-Operators/Tour-tech

Batten, C. L. (1978). *Pleasurable Instruction: Form and Convention in Eighteenth-Century Travel Literature* (Vol. 6). Berkeley: University of California Press.

Brito, L. M. & Carvalho, C. (2021). Reflections on how the Covid-19 pandemic can change tour guiding. *International Journal of Tour Guiding Research, 2*(i), 1-2. Retrieved from https://arrow.tudublin.ie/ijtgr/vol2/iss1/2

Brito, L. M., & Farrugia, G. (2020). On tourist guiding: Reflecting on a centuries-old profession and proposing future challenges. *International Journal of Tour Guiding Research, 1*(1), 4-12. Retrieved from https://arrow.tudublin.ie/cgi/viewcontent.cgi?article=1007&context=ijtgr

CEN. (2003). *What is a Tourist Guide?* Retrieved November 15, 2021, from WFTGA: https://wftga.org/about-us/what-is-a-tourist-guide/

Cetin, G., & Yarcan, S. (2017). The professional relationship between tour guides and tour operators. *Scandinavian Journal of Hospitality and Tourism, 17*(4), 345-357.

Chambers, E. (2001). *Native Tours: The anthropology of travel and tourism.* Prospect Heights: Waveland Press.

Cohen, E. (1985) The tourist guide: the origins, structure and dynamics of a role, *Annals of Tourism Research, 12*, 5-29. https://doi.org/10.1016/0160-7383(85)90037-4

Dewar, K. (2000). An incomplete history of interpretation from the big bang. *International Journal of Heritage Studies, 6*(2), 175-180.

Gebicki, M. (2018). *Do guides get a commission when they take you shopping?* Retrieved August 28, 2021, from Traveller: https://www.traveller.com.au/do-guides-get-a-commission-when-they-take-you-shopping-h11q3p

Goyal, A. (2020). *How did they travel in ancient India? – Sumedha Verma Ojha.* Retrieved November 15, 2021, from Inditales: https://www.inditales.com/travel-in-ancient-india-sumedha-verma-ojha/

Gurung, G., Simmons, D., & Devlin, P. (1996). The evolving role of tourist guides: the Nepali experience. In R. Butler, & T. Hinch, *Tourism and Indigenous Peoples* (pp. 107-128). London: International Thomson Business Press.

IATO. (2019). *Agreement between IATO and TGFI.* Retrieved October 15, 2021, from IATO: https://iato.in/uploads/pdf/GuideFeefrom1stOctoberto30thSept2021.pdf

Jezewski, M., & Sotnik, P. (2001). *The Rehabilitation Service Provider as Cultural Broker: Providing Culturally Competent Service to Foreign Born Persons.* Buffalo, NY: Center for

International Rehabilitation Research Information and Exchange.

Kainthola, S., Tiwari, P., & Chowdhary, N. R. (2021). Tourist guides' perspectives of demarketing the Taj Mahal. *International Journal of Tour Guiding Research, 2*(1), pp. 17-30. Retrieved from https://arrow.tudublin.ie/ijtgr/vol2/iss1/4/

Morgan-Proux, C., & Cable, T. T. (2018). *The Nature Guides of Chamonix.* Legacy, National Association for Interpretation. Retrieved October 23, 2021, from The HAL Portal of Clermont Auvergne: https://hal.uca.fr/hal-03024855/document

Murali, K. (2017). *Conundrum of Tour Commissions.* Retrieved August 28, 2021, from Chronicles of a World Traveller: http://www.anasiantraveller.com/tour-commissions-conundrum/

NZ History. (n.d.). *The Guides of Whakarewarewa.* Retrieved November 10, 2021, from New Zealand History: https://nzhistory.govt.nz/women-together/guides-whakarewarewa

OCED. (2020). *Tourism Policy Responses to Coronavirus (Covid-19).* Retrieved September 3, 2021, from OECD Policy Responses to Coronavirus (Covid-19): https://www.oecd.org/coronavirus/policy-responses/tourism-policy-responses-to-the-coronavirus-covid-19-6466aa20/#back-endnotea0z56

Pond, K. (1993). *The Professional Guide: Dynamics of our guiding.* New York: Van Nostrand Reinhold.

Reed, T. (2015). *Can apps replace tourist guides?* Retrieved April 24, 2021, from Daily Tourism Update: https://www.tourismupdate.co.za/article/can-apps-replace-tourist-guides

Reisinger, Y., & Steiner, C. (2006). Reconceptualising interpretation: the role of tour guides in authentic tourism. *Current Issues in Tourism, 9*(6), 481-498.

Sharma, S., Chowdhary, N., Prakash, M., & Devrath, R. (2014). *A Textbook of Travel and Tourism.* New Delhi: Goyal Brothers Prakashan.

Weiler, B., & Black, R. (2014). *Tour Guiding Research: Insights, Issues and Implications (Aspects of Tourism, 62).* Channel View Publications.

Weiler, B., & Black, R. (2015). *Tour Guiding Research: Insights, Issues and Implications.* Bristol, England: Channel View.

2 The Tourist

Learning objectives

After studying this chapter, you should be able to understand:

- Who is a tourist and who matters to a tour guide
- The different types of tourists and how they relate to tour guides
- The different stages of travel and visitor concerns, especially when they are arriving from a different culture
- What the guests expect a tour guide to do
- The concerns of tourists with disabilities. How can tour guides help them?
- Who is a responsible tourist, and what is expected of a tour guide?

The protagonist in this book is the professional who accompanies the visitors to a site and helps them make sense of the place and activities. In common parlance, this visitor is called a tourist, and this professional is referred to as a tour guide. However, in the strictest definitions, the customers of a tour guide are both tourists and day visitors. As a service provider, tour guides should perform to meet the expectations of their consumers. Therefore, they must understand the underlying consumer behaviour and the necessary distinctions. In this chapter, we discuss the consumers. We refer to them as tourists, including all visitors to the place. We also discuss the tourist concerns. Increasingly, and especially after Covid 19 pandemic, visitors are becoming aware of the need to be responsible. We, therefore, devote a discussion to responsible tourists. There is a realisation that tourism must not leave anyone out of its ambit. A section is therefore devoted to tourists with disabilities.

The tourist

A popular definition of tourism is that it is the activities of the tourist. So, let us begin with an understanding of the term 'tourist'. According to International Recommendations for Tourism Statistics 2008 (United Nations, 2010), a visitor is a traveller taking a trip to the destination outside his/her usual environment, for less than a year, for any primary purpose (business, leisure, or other personal purposes) other than to be employed by a resident entity in the country or place visited. These trips taken by visitors qualify as tourism trips. Thus, tourism refers to the activity of visitors.

Further, a visitor (domestic, inbound, or outbound) is classified as a tourist (or overnight visitor) if his/her trip includes an overnight stay or as a same-day visitor (or excursionist) otherwise.

Figure 2.1: Definition of tourist (United Nations, 2010, 2.13)

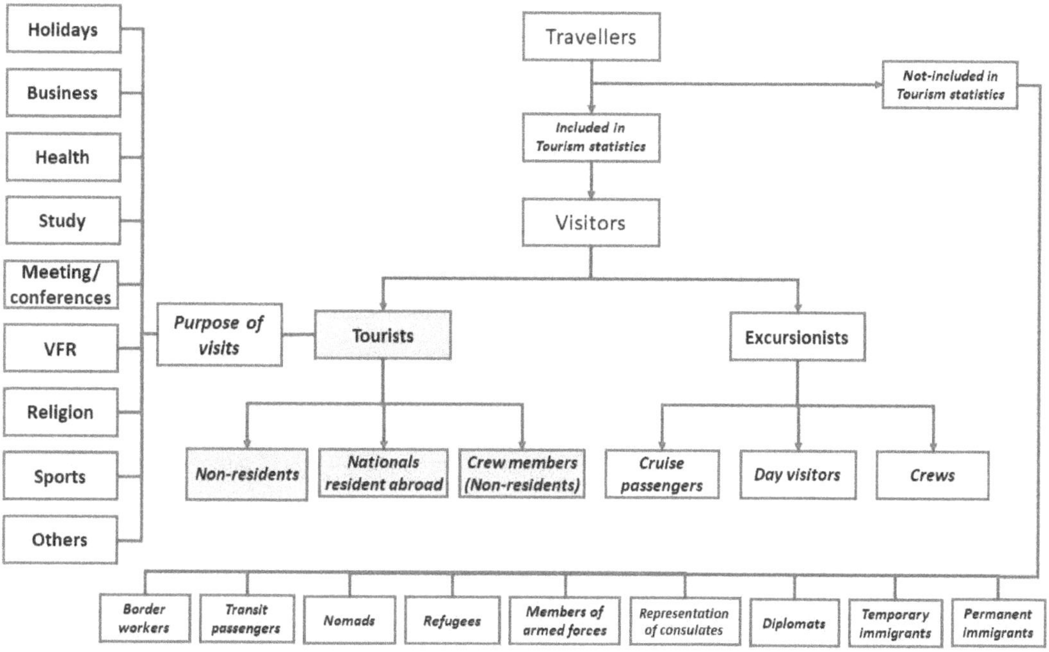

Source: Adapted from Holloway (1989)

Vasudevan, Vijayakumar, & Saroop Roy (2017) summarised the above into two: who are included as tourists and the travellers not included as tourists.

Tourists will include:

People travelling for leisure, health, or VFR (including nationals who live permanently abroad)
Participants travelling to attend meetings or for assignments (sports and management)
Expatriate employees on assignment abroad for less than one year
Business travellers (short assignments)
Those studying abroad- students and young people at boarding schools or colleges and those who travel or work temporarily during their holidays
Visitors from cruise ships, even if their stay is less than 24 hours
Transit passengers who cross the country for more than 24 hours
Foreign airline and ship crew on a stopover in a country
Musicians and artists on tour

Travellers who are not tourists include:

Diplomats and representatives of consulates (travelling for duty purposes)
Members of the armed forces
Refugees (as defined by the UN Commissioner for Refugees, 1967)
Transit passengers (who do not leave the transit area)
Nomads (as defined by the UN in the statistics of international migration, 1980)
Permanent and temporary immigrants
Border workers

Types of tourism

According to Eurostat (2014), tourism can be classified into the following three types based on the origin and destination of the visitors.

- **Inbound tourism** – Visits to a country by a non-resident of that country- for example, when a Chinese citizen, Mr Zheng, comes to France to see the Eifel Tower, he is an inbound tourist for France.
- **Outbound tourism** – Visits by the residents of a country to another country, for example, when a French citizen, Jean, goes to Singapore, he is an outbound tourist for France.
- **Domestic tourism** – Involves travelling by the residents of the given country within their own country, for example, when Jacques, a resident of Normandy, goes to visit Eifel Tower in Paris, he is a domestic French tourist. Of late, in larger countries, travellers from within the province or region are often referred to as local tourists. A tour guide needs to appreciate the expectations of domestic visitors as distinct from the local visitors in terms of language or even cultural differences. Technically, local tourists are included in the broader category of domestic tourists.

Based on the above definitions, we can have the following regrouping:

- **Internal tourism** comprises domestic and inbound tourism. This refers to tourism at a destination with visitors from within the place or outside.
- **International tourism** comprises inbound and outbound tourism. This refers to tourism where the national borders are crossed, whether inwards or outwards.
- **National tourism** includes domestic and outbound tourism. This refers to tourism by nationals of a country, whether they travel within the country or beyond.

FITs and GITs

FIT stands for Free Independent Travellers (or Tourists). It means a small number of tourists (usually fewer than five persons or couples) who take tours at a time of their own choice and are not bound in a group/ groups. They avoid mass tourism and the holiday package concept promoted by travel operators in favour of a more individualistic approach to travel. They search for travel information to learn about attractions, activities and travel routes in the destination. After that, they plan their programme and purchase their flight tickets, hotel, restaurants and transportation reservation. The Internet and mobile technology empower them. Thus, in practice, travellers who have booked only air travel tickets are considered FITs. Backpackers are an important category of FITs.

On the contrary, GIT stands for Group Inclusive Travel. It is a travel program with a special fare and specific requirements (generally a minimum number of persons, often more than five persons travelling as a group throughout the tour). Usually, they travel on scheduled transportation with an intermediary travel agent. Mak (2004) refers to this as Group Package Tour (GPT). GPT includes elements of a leisure trip, such as transportation, transfers, baggage handling, accommodations, sightseeing schedule and other services, into a package sold to the consumer at a single price.

GPT companies purchase services from vendors in bulk and therefore can negotiate a better price. They then pass part of this volume-pricing advantage to their clients. Also, in many cases, they have access and priority to destinations otherwise out-of-reach and expensive to visit. These advantages often offset the self-booking advantage that the Internet offers. Thus, the rate for FIT is often higher than GIT. Although travel companies offer GIT, more often than not they use the services of tour guides. Thus, GIT is an institutionalised source of business for tour guides.

Bonn et al. (2010) point out that the presence of tour guides during the GPT trip make GPT guests feel secure and allows for swifter service recovery when needed. In addition, they observe that group traveller's overall satisfaction was significantly higher than that of the independent traveller. However, it has been reported that independent travel that allows customised experience is rising while group travel offering standardised itinerary and experiences is declining (Katawandee & Slisa, 2014; Chesshyre, 2002; Pryor, 2001; Scutte, 1997).

FIT can also engage tour guides as they arrive at a destination or a monument. In addition, FITs may subscribe to walking tours at the destination. Finally, FITs generate more tourism revenues (Katawandee & Slisa, 2014), and therefore destinations are focusing on them. As a result, travel companies are reassessing their strategies to target this segment. Tour guides must remain aware of this change and reorganise their strategy accordingly.

Travel stages and visitor concerns

A tour guide, to be effective, should understand the challenges and concerns of tourists as they complete their travel. A tourist typically travels through four stages. The characteristics of each stage are as follows:

Arrival: As tourists arrive at a destination, they are confronted with travel-related challenges which bring them both anxiety of the unknown and make them eager for new encounters. As the visitors arrive, they might have jet lag, have to negotiate the immigration, have the first encounters (moments of truth) with the place and the hosts, and tackle the challenges of navigating through the destination. So, guests encounter cultural bumps. A tour guide must meet and greet them, empathise with the guests and facilitate them as they settle at the destination. The guide must be prepared for possible scenarios and have feasible solutions.

Stay: This is the most crucial stage where the guests are likely to spend most of their trip time. They look for attractions, accommodation and activities during their stay. They also seek amenities to make their stay comfortable. During their stay, they want the promised services. They seek information as they go around the place. They expect the tour leader to arrange for all these and remain present to attend to situations as they arise.

Departure: Guests expect tour leaders to support their transfer to their departure stations – airports, seaports, bus stands, or train stations, whatever the case may be. Tourists try to hold on to the experiences by firming up the last impressions as they travel back home.

Back to home: On return from holiday, visitors would like to cherish memories of their visit. They would like to have pictures and souvenirs for reminiscence. Moreover, visitors would like to record their complaints in case of inadequate services. Even this should be facilitated. Tour guides sometimes remain in touch with their guests until long after the trip and encourage them to post good reviews and to travel again with the company.

Visitor concerns

Usual visitor concerns during travel are related to the following. A tour guide must understand and anticipate these and try to address these.

Transportation: Transport should be safe and reliable. If the signage at the destination is unorganised or the luggage goes missing, they will need help. They also need advice on finding stations, locating transport and local tricks of maximising their returns in terms of efforts and cost.

Lodging: Visitors judge accommodation according to cleanliness, location (accessible vs exotic), and services. They want tour leaders to help them settle in, sometimes help them register their complaints and concerns with property managers, and if that does not work, help them find alternative accommodation.

Dining: Food is a significant concern of many visitors. While some are sceptical about trying out new things, others are more adventurous and want to experiment with local cuisines. Some guests may have concerns about hygiene or need vegetarian or halal or kosher food. Therefore, they want tour guides to advise them on food quality, price, local speciality food, and local liquor.

Shopping: Visitors want to carry back home memories. They would like tour guides to advise them about a good place to shop for souvenirs, artefacts, and collectables, where they find a variety and genuine products and at reasonable prices. They also want guides to advise them if goods are prohibited from taking out of the destination and subject to restrictions imposed by airlines. For example, India does not permit certain goods of social, health, environmental, wildlife, and security concerns. An example of this could be old Indian coins covered by the Antique and Art Treasure Act, 1972. Similarly, New Zealand subscribes to the Convention on International Trade in Endangered Species (CITES) to prevent endangered, threatened or exploited species from being taken out of the country.

Recreation and **entertainment**: Visitors seek advice and assistance to participate in special interest activities, cultural experiences, and immersion in the host culture. This is where the tour guides have an opportunity to impress their patrons and engage them in additional activities.

Security: Safety is a significant concern of visitors. There are travel advisories. However, visitors do not take them at face value. Once at the destination, guests want tour leaders to advise them about it. What is a safe area? What is a safe time to be out? What kind of petty crimes should one be vigilant of? They also seek advice on safety protocols for participating in activities, especially adventure.

Expenses: People generally look for value for their money. On the ground, guides may be the local advisor of shopping, buying food and tricks to save money on local things, including transportation. In some countries, it may be possible for the tour guides to use their network to get a good bargain for the visitors.

Responsible conduct: Guests are increasingly conscious of their behaviour. They want to be ethical and responsible in addressing local sensitivities. They seek advice on protocols for participating in local activities where some norms or rituals may be culturally sensitive. We discuss responsible tourists separately later in this chapter.

Challenges faced by international visitors

A tour guide must not consider every issue of a visitor as a symptom of cultural shock, but they must understand and empathise with some of the genuine concerns of visitors, especially the international tourists. Some of the common problems faced by international tourists are:

Jet lag

Jet lag, medically referred to as *desynchronosis*, is a physiological condition that results from alterations to the body's circadian rhythms resulting from rapid long-distance transmeridian (east-west or west-east) travel on a jet aeroplane. It is a sleep disorder. The condition of jet lag may last several days until one is fully adjusted to the new time zone, and a recovery rate of one day per time zone crossed is a suggested guideline. However, tourists may not get enough time to recover from jet lag. Therefore, jet lag is best managed during travel. One option to counteract jet lag is to break the trip into smaller segments if it is too long and stay overnight at an intermediate destination.

It may be advisable to adjust sleeping hours on the plane to match the destination time. However, this may not always be possible for the visitor. Additionally, sometimes they are too nervous to sleep on the aircraft. Therefore, the best thing to do is adapt to the local time and eat accordingly upon arrival. Also, exposure to sunlight during the day is helpful. Finally, a tour guide must empathise with guests who might report jet lag.

Weather

Another substantial change that may be disruptive for tourists is a sudden weather change. Though a change in weather could motivate the holiday, coping with it can be challenging. Extreme heat or cold or rain and snow may take tourists by surprise if they are not fully prepared and are not advised for the occasion. High altitude (low pressure) and heat (drying up) may slow tourists. It may be extra pressure on their heart, so they must be advised not to make haste and take time to acclimatise. Guides must also not rush them through the itinerary during this phase. Instead, they must be cautioned to take time and not go at their usual pace. For example, in Ladakh, a complete rest day is advised on arrival for acclimation due to high altitude.

Language

Language differences could annoy if other things are not organised properly. For example, buying tea or coffee or making a purchase could be a fun experience for some but could irritate many if they have problems understanding transactions. A tour guide must help visitors with basic salutations and courtesies, and foreign language skills will be an excellent advantage. The guides can also advise restaurants and markets where the visitors' language is spoken.

Food

Local food is a source of excitement for many tourists. However, it is also the most common source of the complaint. Food in India is typically spicy and hot. It may taste great but may cause the disorder. On the other hand, good food is the fuel for the visitors to keep the interest growing through the day, so the appetite must be addressed. Adding to the challenge are different timings for dining, menus in a strange language, various options that tourists do not understand, and personal reservations for a particular type of food (vegetarian for many Indian, or kosher for Jews). Encourage your guests to be more adventurous with food. However, remember they might ask for a simple food (fruits or boiled vegetables or only a pizza) some time. In such a situation, guides must also be equipped with options closer to the tourists' taste buds.

Illness

Nothing is more upsetting for the group than to have one of the members falling ill. They fear their limited time being diverted to taking care of the member in an unfamiliar place. A foreign tourist is more likely to be unwell because of cultural shock or food or a change in the climate. The guide should nip the problem in the bud, else it will worsen and may slow down the tour. Call for immediate medical help. Inform the tour company and advise guests to rest rather than exert themselves. Tour guides are recommended not to try their hand at medication. They must seek expert advice.

Unfit tourist

Tourists arrive at their destination with lots of dreams and expectations. They often underestimate their fitness to travel and undertake a tour to a demanding attraction. For example, a tourist with heart disorder comes to a point where there is a steep trek to walk, so he cannot go further. This situation is a real misfortune and is a sad state. Guides should not insist. See if they could be helped with a partial walk.

Overindulgence

Some guests are foodies and imbibers. Their idea of a holiday is overindulgence. If not restricted, effects would be felt sooner or later on tour. A tour guide's responsibility is to keep advising guests about the demanding schedules ahead and that they might miss out on exciting experiences should they overindulge.

Loss of personal property

This issue is the most common source of wasted time for the group. Unfortunately, not much can be done. However, allow extra time (up to many hours) for lost passports, arrange for tickets, and file claims for lost and 'not arrived' baggage. Typically a client with such issues may be disengaged from the tour party, and someone else from the tour company takes care of these matters of this guest.

Disproportionate expectations

A tourist is often disappointed by unmet expectations. Expectations might be a figment of their imagination based on social media, or an overzealous salesperson might have tantalised them. It is advised that tour guides carry necessary documents/ tour brochures to set the expectations right. Please explain politely what is included in the tour and what to expect.

Why do visitors need a tour guide?

As soon as possible, a tour guide must determine the guest's/group's expectations. Then, they must prepare to conduct the tour accordingly. Generally speaking, some of the reasons why people hire tour guides include:

- To ***know*** the destination's history briefly and pointedly, geography, society, culture, traditions, everyday life, etc.
- To ***understand*** the foreign culture which is different to theirs. Tourists are visiting the place (perhaps) for the first time and would like the tour guide to explain all the exciting things that they come across.
- To suggest and ***provide access*** to the otherwise non-public territory. Visitors in unknown places are sceptical in traversing the unknown but possibly exciting area. Guides are their pathfinders and pilot.
- To ***explore*** new places, areas, and phenomenon
- To ***suggest*** an itinerary (what to see and experience and what not to see) depending on the guests' taste, time, and resources.
- To ***help*** with managing the limited time that visitors have at a destination. Guides help visitors by taking them through more significant and exciting points in the shortest possible time and dividing available time among these attractions and utilities.
- To ***benefit*** from expertise to be introduced to an area by someone with detailed information and personal experience.
- To ***enjoy*** being taken care of. Guests would like to leave their samey worries like tickets, reservations, time management, vehicle arrangement, and similar things to someone else and enjoy the tour.
- To ***avoid*** the fear of being lost in an unknown place.
- To be ***able*** to meet and mix with natives without offending their sensitivities.
- To ***try*** their hand at local things and have the excitement (do not underestimate the importance of `being part of the crowd '). However, of course, they expect tour guides to help.
- To ***travel*** without inhibitions and anxiety of being in an unknown land.

Tourists with disabilities

The World Tourism Day 2021 was themed 'Tourism for inclusive growth'. The vision of inclusivity encompassed the idea of leaving nobody out of tourism – even the differently-abled individuals; that tourism is enjoyed widely and fairly. Végh (2005) suggests that everyone should be able to enjoy travel because enjoying life to the full is everybody's right.

Goal 11 of the Sustainable Development Goals (SDGs), a part of the 2030 Agenda for Global Action, is to make cities inclusive, safe, resilient and sustainable. This goal includes tourism and recreation is its call for universal design for accessible and sustainable transport systems, inclusive urbanisation, and access to green and public spaces.

Of late, it has been argued that destinations need to offer touristic experiences beyond mere visual gaze. Small, Darcy, & Packer (2012) and Efstathiou & Zippelius (2019) argue that such experiences should tantalise not just sight but also sound, taste, feel and smell, encapsulated by sensoryscapes. Furthermore, the UN Convention on the Rights of Persons with Disabilities, adopted in 2006, requires the state to ensure accessibility for people with disabilities to sporting, holiday, and tourist venues and services (Act XCII of 2007).

The term 'disability' has a broader meaning and scope. Conventionally it is viewed as reduced mobility or visual, hearing, and intellectual impairments. Others include long-term illnesses that affect persons' quality of life, also as disabilities (Zsarnóczky, 2018). It is estimated that because of a variety of social conditions, there will be around 1.2 billion persons with disabilities (PwD) by 2050 (World Health Organisation and World Bank, 2011). Spouses, children and caregivers often accompany PwDs and counting them in, almost a third of the world's population is affected by disabilities.

However, disability may be short term or temporary, for example, pregnant women and people rendered less mobile due to accidents. In addition, many of us face increased limitations in travelling as we grow old. Modern society's ageing profile and pattern suggest a continuous increase in people with age-related disabilities. They face similar limitations and impediments and have needs similar to people with disabilities (Figueiredo et al., 2012).

It is often assumed that PwDs are often not interested in travelling (Loi & Kong, 2015). However, PwDs have the same tourism desires as other groups (Yau et al., 2004). Thus, there has been a regular call for attention to this segment over time. Attending to this segment of visitors is viewed differently by stakeholders. For some, it is a cause, and the principle of inclusivity mandates that they are accessible to all. For others, it is a distinctive feature that creates market opportunity. Therefore, whether private or public, destinations are making efforts to make themselves accessible to all regardless of physical barrier, disability, or age. However, the presence of different disabilities requires that these are attended differently. For example, the needs and requirements of visually challenged people will be quite different from those with hearing impairment or those with mobility issues. Further, people with the same disability may have different levels of functioning.

According to the European Network for Accessible Tourism (ENAT), all components of tourism should facilitate persons with disabilities. Furthermore, Article 2 of the Convention on the Rights of Persons with Disabilities (CRPD) also states that the destinations should use disability-inclusive universal design. 'Universal design' means the design of products, infrastructure, facilities and services to be usable by all people, to the greatest extent possible, without the need for adaptation or specialised design, which does not exclude assistive devices for particular groups of persons with disabilities where this is needed. Thus barrier-free destinations are conceived.

Transport by land, sea, and air should be suitable for all. Furthermore, activities, exhibits and attractions must be designed to allow participation of PwDs, and the information offered in booking systems, websites, marketing material and other services should be accessible to all. Last but not least, some tourists with handicaps will require assistance from highly skilled staff.

Defining a tourist with access needs

A range of terms is used somewhat loosely to mean people with disabilities. Phrases like people with disabilities (PwDs), tourists with disabilities, tourists with access needs, tourists with special needs, differently-abled, etc. are used interchangeably to mean the same thing. Researchers in disability language refer to the different schools or models the users subscribe to. Though there are numerous models available, some basic ones include medical, charity, social and human rights schools.

In the past, the individual or medical models had used terms like 'handicapped', 'spastic', and 'retard', which were considered hurtful. Terms used with good intention may also be painful to the individual, as in the social and UK model, which use the term 'disabled person', or in the human rights (UN) model, 'person with disabilities' (Gillovic & McIntosh, 2015). In the same vein, sometimes 'differently-abled persons' or 'specially-abled persons' is used. However, a review of the academic literature suggests that these terms are scantily used.

There are harmonised guidelines and space standards for the barrier-free built environment for persons with disabilities and elderly persons (Ministry of Urban Development, Government of India, 2016). These refer to 'Persons with Reduced Mobility' (PRM) and 'Persons with Disabilities' (PwDs). Aviation vocabulary refers to travellers with a disability as 'Passengers with Reduced Mobility' (PRMs) (European Regions Airline Association, 2018). There are many instances cited where PRMs are mistreated, and at times airlines have been fined for such conduct. Many airlines define PRMs who might be unable to walk long distances, ascend stairs, or be completely immobile. Accordingly, civil aviation authorities create a legal entitlement called 'special assistance' for PRMs travelling by air – for example, the EU Regulation (EC) 1107/2006.

Expectations and challenges

One of the crucial touchpoints for the visitors, including guests with disabilities at a tourist site, is a tour guide who mediates between the guests and the attraction. According to Zhang & Chow (2004), tour guides are frontline personnel who help create the 'moment of truth' for the guests and therefore have a significant role in making or breaking the trip experience. Earlier, Moscarado (1998) identified three ways tour guides contribute to a guest's experience. One, they provide information on available options so that a tourist can make the best choices about what they do and where to go. Two, they provide information to ensure that guests are safe and comfortable. These inputs help them know how to cope with and better manage encountered difficulties and understand messages given by the warning signs. Three, they help create the experience so that guests can participate in heritage walks, adventure, visit art galleries, and learn in areas of educational interest. In the case of guests with disabilities, tour guides have additional responsibilities.

Among the challenges for tour guides is balancing between creating the desired experiences for PwDs and their ability to consume the experience in a manner available for the general public. For instance, concern for their safety can be genuine and therefore, the accompanying tour guide may have to cite those limitations, or the site authorities may not have the requisite infrastructure to address the access needs of the PwD segment of visitors. Further, Burns, Paterson, & Watson (2009) note that PwDs are a varied group in terms of experiences, views and needs, and therefore, these guests with different disabilities have different information requirements. It is therefore challenging to offer customised information and services for them. It is generally seen that the mobility challenge is addressed first and is often the only concern addressed by service providers like museums and galleries, archaeological sites and parks. Even the accommodation providers fail to facilitate the PwDs with other impairments.

Sites and attractions often do not have skilled tour guides who can take care of people with disabilities, especially those with sight and hearing impairments, to enable them to enjoy tourism activities. Thus, the site authorities could provide a brochure or leaflet with all necessary information about the attraction in the case of hard-of-hearing guests. In addition, the tour operators should also provide a tour guide skilled in sign language. The absence of these inclusions would mean offering hearing-impaired guests lesser experience than the others.

Depending upon the targeted segment of PwDs, the tour operators may decide to offer a range of services for barrier-free holidays, including attendant care, and specially designed accessible vehicles for transfers and transport. They may also provide services like shower and dressing, assistance in getting on and off the vehicles, and movement up/down the stairs. Often the tour guides are trained for these responsibilities.

Further, the site authorities may identify the routes with access. The tour operator, on their part, may design itineraries that include such sites that offer access to PwDs. They may create and share information about such itineraries to enable guests with disabilities to choose suitable tours. Their brochures and websites may include photo galleries of facilities like wheelchair accessible transport, entryways, toilets, and accommodation for guests with disabilities. Information about situations where access is an issue must be shared, and how such situations are handled may also be communicated.

Guests with a disability expect that the tour guide understands what the tour operator had promised and is sensitised about the challenges they face. Tour guides must work with the tour companies to curate a range of activities and experiences for guests with a disability. They expect that a professional tour guide works out the details like scheduling convenient departures and arrivals; connecting with transporters who provide accessible vehicles and ensure that they are in order; coordinating with attendants at the site if required; organising appropriate equipment, and paraphernalia for them.

Tour guides, on their part, have reported challenges that they face accompanying guests with access needs. These include issues related to time management, language selection, equipment barrier and environment barrier (Chikuta et al., 2017). The *equipment barrier* refers to equipment made available to tour guides by site authorities or tour operators to facilitate the guests with special needs, where such equipment is inadequate or inappropriate for their needs.

The *environmental barrier* refers to limitations on the movement of guests within the servicescape. On the one hand, the challenge is that the site does not have accessible pathways (ramps, staircase, lift, corridors, etc.) and signage. On the other hand, creating such a facility may ruin the visual gaze, spoiling the place's authenticity and appearing too artificial.

Inexperienced and untrained tour guides also face challenges in communicating, interpreting and sharing information with disabled guests. In addition, they are often forced to limit their tour to activities they deem safe for these guests, notwithstanding what the tourists want. For example, game park tours require long walks or drives. Often these have to be curtailed.

As guests with disabilities and their companions emerge as a significant market, tour guides, supported by the site authorities and tour operators, need to skill up to handle such visitors. To begin with, they must be aware of obligations set out in the Convention on the Rights of Persons with Disabilities (OHCHR, 2006). Then, they must be trained to capacitate them to handle and treat well tourists with disabilities. Such training may comprise appropriate communication skills, including the basic sign language, handling and care procedures of different disabilities, and health concerns of the tourists with disabilities (Chikuta et al., 2017).

Besides deploying universal design principles, site authorities should also use technology. For example, they should provide information in braille, tactile graphics, overlays, interactive interfaces, and audio descriptions for visually impaired visitors. In addition, experiences can be enhanced with tactile 3-D printed reproductions as gamification may be included to awaken other senses and maximise the enjoyment of the artwork. Furthermore, tour guides should be conversant with these technologies and help guests benefit from them, and the relevant authorities should train tour guides to use the technology and the equipment. Similarly, they must be trained to help visitors board and descend from specially designed vehicles. In addition, they must be aware of the facilities like restrooms, pavements, entrance and exits and rooms for disabled guests at the site. They should also have access to the equipment for the tourists with disabilities like wheelchairs, crutches, walking sticks and portable ramps.

Responsible tourism

Post-Covid-19, it is observed and reported that travellers have become more concerned and are behaving more responsibly. Today, it is widely accepted that tourism impacts and therefore, it must be organised and consumed responsibly. The term 'responsible tourism' first appeared in the literature in the early 1990s, in Smith's report on the 1989 World Tourism Organisation-convened seminar on 'alternative' tourism in Tamanrasset, Algeria. Formally, responsible tourism was recognised and accepted at World Summit on Sustainable Development in 2002. It is also referred to as the Cape Town Declaration.

According to (Goodwin, n.d.), Responsible Tourism (RT) is about "making better places for people to live in and better places for people to visit." To make tourism more sustainable, RT puts obligations on all stakeholders – operators, hoteliers, governments, local people, and especially the tourists. Therefore, responsible tourism, on the one hand, aims to reduce negative economic, environmental, and social impacts, while on the other hand, it boosts local economies and improves the wellbeing of host communities. It contributes to preserving natural and cultural heritage while providing tourists with more enjoyable experiences through meaningful connections with local people. Responsible Tourism creates a better understanding of local culture, social, and environmental issues; provides access for physically challenged people; and is culturally sensitive, engendering respect between tourists and hosts, and building local pride and confidence.

As humans, we are a part of the ecosystem. According to scientists, an ecosystem is an community of biotic and abiotic components linked together through nutrient cycles and energy flows during continuous interactions. Humans exist vis-à-vis the living systems like animals, plants, soil, air, water, and other elements that support them. However, even though a subtle equilibrium exists, human activities, including tourism, tend to disturb this balance. Tourism can lead to negative consequences on the environment, thereby reducing the quality of life of locals and adversely impacting the experience of the visitors (Muhanna, 2006; Kainthola et al., 2021). Thus, responsible tourism also has environmental, economic, and social dimensions (Booyens & Rogerson, 2016). Therefore, responsible tourism has been proposed as an approach to maximise the positive and minimise the adverse effects of tourism.

Sustainability is considered as minimising the damage caused due to human activity, which implies the consumption of resources. The fundamental assumption is inherently contradictory. On the one hand, economic activities are considered essential and economically gainful. On the other hand, it is professed that such activities should be efficient and minimise the impairment caused therein. Thus, while tourists consume resources at a destination, they contribute economically. However, after the pandemic, more sensitised travellers question how their travel will make the place better. The answer lies in balancing the consumption patterns and the contribution made. It is not only about abandoning a polybag; it is also about asking who grows the crop, who prepares the meal, and how they will benefit from my travel. It is about conscious decisions to help the host community and the local ecosystem. It is about giving more to the community than just monetary benefits.

Responsible tourist behaviour

For a long time, the responsibilities of tourists have been discussed. Krippendorf (1984) describes a responsible tourist as "one who chooses those forms of travel which are least harmful to the environment, which are least disturbing for the people and cultures of the tourist areas and from which they get the greatest benefit. He spends his money on those products and services about which he knows the origin and who will profit from their spending. He observes these principles when choosing accommodation, food, means of transport, visiting institutions, buying souvenirs. He takes time to prepare his journey, and he stays as long as possible in the places he visits so that the experience may be a lasting one and that he may really identify with it."

Thus, the tourists are responsible for their conduct and consumption. According to Swarbrooke (1999), these responsibilities include:

- Obeying local laws and regulations;
- Not participating in activities widely condemned by society, even if they are not illegal or where the local authorities do not enforce the law;
- Not purposefully infringing on local religious beliefs or cultural norms of conduct;
- Not deliberately harming the local physical environment;
- To minimise the use of scarce local resources.

Theory of mindfulness

The best way to understand mindfulness is to contrast it with mindlessness. Mindlessness is conducting oneself as per routine without the application of consciousness resulting in an automatic response to a situation. Constraining oneself to mundane rules and regulations without questioning is another example of mindlessness. Context governs behaviour, but mindset governs how the context is interpreted (Stanford, 2006). For example, visitors frequently come from culturally diverse backgrounds. A visitor who is less aware of their surroundings will not try to reconsider how this new context should be interpreted.

On the other hand, mindfulness is the fundamental human ability to be fully present and aware of where and what one is doing (Mindful, 2021). According to Langer (1993), "mindfulness is a state of mind that results from drawing novel distinctions, examining information from new perspectives, and being sensitive to context…When we are mindful, we recognise that there is not a single optimal perspective, but many possible perspectives on the same situation."

Moscardo (1996) believes that interpretation makes visitors more mindful so that they are active, interested, questioning and capable of reassessing the way they view the world. A varied, multi-sensory, vivid interpretation stimulates visitors to encounter a new context or exhibit. Actually, the unexpected context rouses heedful inquiry that helps produce a mindful response to what exists in the surroundings and what is being interpreted. Moscardo (1996) concludes that such mindful visitors realise the consequences of their actions and will therefore conduct in a manner that lessens their impacts on a given site and community.

Therefore, the theory of mindfulness helps make sense of why a visitor will behave responsibly or not. The pre-Covid visitors were sometimes casual in their approach, conducting themselves out of habit because that is the way they have been all through, and there was not much reason to reassess this habit. So, they mindlessly applied the values and attitudes they had travelled with earlier without questioning whether or not this was appropriate. However, Covid has had a runaway effect on our thinking. It has shown how the planet became a better place with restricted human activity. As a result, the rivers got clean, the mountains were tranquil, forests were serene, and even the skies over the cities were clear and bright. Tourists now realise that their negative impact on the destinations would be limited if they conduct themselves with more discipline.

Tour guides have a significant role in educating visitors for responsible conduct and promoting sustainability. One, they motivate visitors to modify their consumption behaviour and therefore restrict the adverse impact on the site. Two, they are the environmental interpreters who foster understanding and appreciation of environmental issues to facilitate responsible tourist behaviour in the long term (Weiler & Ham, 2002).

Tour guides have extended contact with the visitors and interpret the destinations' offerings, laws, rules, regulations, expected behaviour (Kabii & Wandaka, 2017; Sandaruwani & Gnanapala, 2017). Interpretation by tour guides can relieve crowding and congestion, alter behaviour directly by giving information or fostering appreciation, and create public support for conservation (Moscardo, 1996). Tour guides, therefore, have a role in promoting sustainable tourism and affirmatively influencing responsible behaviour at the destination.

Motivation for responsible behaviour

Prior to Covid-19, there was little empirical evidence to appropriate motives for responsible tourist behaviour. However, as early as 1999, researchers, while discussing green tourist motivations, pointed out that the desire to 'feel good' and peer pressures shape responsible tourist behaviour (Swarbrooke & Homer, 1999). Tourists' attitudes, actions and motivations shape this behaviour.

François-Lecompte & Prim-Allaz (2009) have identified five factors to classify responsible tourists into two. The five factors include sacrificing their comfort, the desire to travel with a responsible tour operator, the intention to safeguard local resources, the acceptance not to travel too far, and the desire to protect the cultural and natural heritage of the visited areas. Based on these factors, François-Lecompte & Prim-Allaz (2009) suggest two types of responsible tourists – 'the *sustainable adventurer*' and the '*neo-sustainable*'. Sustainable adventurers are ready to travel sustainably and are the perfect target for tour operators positioned on sustainable development. On the other hand, neo-sustainable responsible tourists are typically ethical travellers who travel close to home to avoid pollution linked to air transport. Further, they avoid tour-operator led organised trips and prefers to arrange for their stay themselves and use the opportunity to visit their relatives. Thus, they are interested in accommodation, leisure, or catering offers close to the inhabitants and nature.

Responsible tourism has become an integral part of lifestyle and ethical marketing. Environmental issues are important to many travellers and holiday-makers when selecting and purchasing travel destinations (Goodwin & Francis, 2003). Further, when tourists care for the heritage, they are willing to pay for sustainable practices as part of their assumed responsibility even when heritage sites' economic value far exceeds monetary benefits (Kim et al., 2007).

One of the challenges is visitors' lack of awareness, compounded by uncertainty whether their actions could make a difference. The reasons include feelings of disempowerment and unwillingness to make significant changes towards prevailing responsible tourism behaviour (Teo et al., 2014). For example, in the case of cultural tourists, the more they perceive the site as part of their heritage, the greater interest they show in learning, feeling emotional involvement and being connected to their heritage (Poria et al., 2006), and therefore believe in being part of and preserving and sharing cultural heritage (Teo et al., 2014).

Regenerative tourism

The underlying assumption is that an ecosystem has an innate ability to recuperate. Sustainability restricts consumption and the resulting impairment to a level lower than the natural recuperation rate of a place. While the *status quo ante* is in our interest, there is an increasing demand for regeneration. Regeneration is an affirmative and conscious action taken for leaving a place better than before. It assumes that the world is nature's living system and not a mechanical one. Consequently, many believe that catastrophes like the Covid result from humans disturbing the ecological equilibrium.

Accordingly, regenerative travel is about the wellbeing of the local ecosystem. Thus, tourism is not just an economic sector; it is dynamic. It is an all-pervasive value system that emphasises regenerative tourism culture. It can be done by designing products and experiences that limit the harm caused to nature and go beyond to restore the damage that we have already done to the natural world. It also requires a colossal attitudinal shift in the demand and supply of tourism. Thus, the role of a tour guide as a mentor and mediator becomes significant.

Regeneration is about the ability to regenerate the land, the biodiversity; empower the local community; and it enables us to create a better experience and satisfaction after contributing to the cause. It takes a holistic view to restore the conditions of life to flourish by resituating humans' wellbeing within the precincts of the wellbeing of the ecosystem. It is about creating a meaningful environment for transformative experiences – beyond the numerical goals of tourist traffic. Regenerative tourism is not about restricting any tourism. Instead, building a regenerative culture is about flourishing, thriving, and striving for wholeness. It is incumbent upon the destination leadership *to create better places to live in and create* great places to live. Everyone can and must contribute. The role of tour guides is vital.

All this while we have approached tourism as a place-based strategy where the superordinate goals were to increase the tourist traffic and revenues. There were some signs of pushback at destinations as the world slipped into the Covid pandemic. Tourism came to a standstill. Nevertheless, while there may be some desperation to bounce back, humans have had their moments of a volte-face. We have had a chance to sit back and rethink – and transcend to a higher level of consciousness with a sense of connectedness with the universe and concern for the greater good. There is a paradigm shift from 'me' to 'we'. There is an increased consciousness that generosity rather than greed, and concern rather than exploitation, are the ways to fulfilment.

Covid-19 has been a watershed in contemporary times, and has had a significant impact on people's worldviews around the globe. The consequences of Covid-19 indicate that this crisis is exceptional and can have a significant and long-term impact on the sustainability of tourism destinations. As a result, tourists have a change in perspective – they are more concerned about the impact of their travel and would like to learn more about the impact of their footprints on the site and contribute to mitigating such impacts. Post-Covid-19, a more thoughtful approach by the tourists can be a key to the steady growth of the travel and tourism sector. A tour guide has a pivotal role to play in this scenario. Through interpretation, they connect the tourists to the place, render them mindful of their impact, advise, and lead them to responsible conduct not only for a sustainable visit but heedful regeneration.

Key points

The protagonist in this book is the tour guide, who accompanies visitors to a site and helps them make sense of the place and activities. However, in this chapter, we will talk about tourists, the consumer of the services of a tour guide.

There are three types of tourism: inbound, outbound and domestic; and tourism can be internal, international, and national. Customers generally purchase tours either as FITs, also called Independent travellers, or GIT the Group inclusive travellers. The latter is also called GPT or Group Package Tours. While GITs are on the decline, empowered by internet, FITs are on the rise.

There are four stages of travel: arrival, stay, departure and back to home. Tourists have different concerns corresponding to each of the stages of travel. These concerns are transportation, lodging, dining, shopping, recreation, security, costs, and responsible behaviour. International guests often complain about jet lag, climate, the local language, food, ill-health and poor fitness, over-indulgence, and disproportionate expectations. Tourists need guides to address their expectations and handle complaints.

This chapter also talks about tourists with disabilities. The vision of inclusivity encompassed the idea of leaving nobody out of tourism – even the differently-abled individuals. The consensus is that everyone should enjoy travel because enjoying life to the full is everybody's right. It advocates the use of principles of universal design. Tour guides need to skill up to handle such visitors.

To begin with, they must be aware of obligations set out in the OHCHR 2006 Convention on the Rights of Persons with Disabilities. They must be trained to handle and treat well tourists with disabilities. Such training may comprise appropriate communication skills. Finally, tour guides should be conversant with technologies to support disabled guests and help them benefit from these.

Responsible tourism is an emerging phenomenon. Tourists have gotten concerned and are behaving more responsibly. They aim to reduce the negative economic, environmental, and social impacts of tourism, while on the other hand, they want tourism to contribute to boosting local economies and improve the wellbeing of host communities. Responsible tourism contributes to preserving natural and cultural heritage while providing more enjoyable experiences through meaningful connections with local people.

Key terms

Tourism The act and process of spending time away from home to pursue recreation, relaxation, and pleasure while using the commercial provision of services (Encyclopedia Britannica). Thus, tourism is a product of modern social arrangements.

Tourist Individuals who travel out of their place of usual residence, for at least one night but not more than one year and not for any remunerated activity in the place to which they travel. For a tour guide, these include day visitors also.

FIT Free Independent Traveller: an individual or a group of fewer than five persons travelling with a self-booked itinerary. This independent travel is not purchased as a package combining transport (air, rail or bus), lodging/ hotel, entertainment, local transport (rental car for destination) or services like a tour guide or a tour manager.

GIT Group Inclusive Travel: purchased as a standard package that includes all expenses such as accommodation, meals, sightseeing transportation, entry tickets, tips, etc. This travel has a fixed itinerary and is operated by the tour operators on a fixed date.

GPT	Group Package Tour: the same as GIT, but includes elements such as transportation, transfers, baggage handling, accommodation, sightseeing schedule and other services, into a package sold to the consumer at a single price.
Cultural shock	The feeling of disorientation manifest in uncertainty, confusion, or anxiety experienced by someone suddenly subjected to an unfamiliar culture and way of life. For example, tourists may face a cultural shock when they travel to another country.
Responsible tourist	One who consumes tourism in a more responsible way that, on the one hand, includes minimisation of negative social, economic and environmental impacts, while on the other generates more significant economic benefits for local people and enhances the wellbeing of host communities.
Tourist with disabilities	A person who has a physical or mental impairment that has a substantial adverse effect on his/her ability to carry out normal day-to-day activities, including travel and tourism. An extended understanding includes the elderly, and pregnant women among others.
Accessible tourism	The ongoing endeavour to ensure destinations, products, and services are accessible to all people, regardless of their physical or intellectual limitations, disabilities or age (UN).

■ Discussion questions

1 Who is a tourist for a tour guide? Talk to some tour guides to determine the difference between a domestic client and an international one. Whom do they prefer and why?

2 It is generally believed that group packaged travel is on the decline. What strategies should tour guides adopt to address this trend?

3 What should be the first things a tour guide should do to assist their clients in recovering from the cultural shock?

4 What are the different stages of cultural shock? What should a tour guide do during each of these stages?

5 Post-Covid visitors have become more conscious of their holiday consumption. Do you agree? How is the tourist changing?

6 Find out if your destination has a responsible tourism guideline. What aspects are challenging for the visitors to abide by? What can be done?

7 Facilitating travel and tourism for disabled visitors is not only righteous, but it also makes good business sense. Elaborate. How should tour guides prepare themselves for leading differently-abled visitors?

■ Activity

1 Talk to some tour guides to understand how FIT business is different from GIT. What are the challenges associated with either?

2 Pair up students from different cultural backgrounds and let them discuss the differences in their culture and societal environments.

References

Bonn, M. A., Chang, H., Agrusta, J., Furr, H. L., & Kim, W. G. (2010). Demographic, behavioral and perceptual comparisons of US visitor experience with Group Package Tours and Free Independent Travel to China. *Hospitality Review, 27*(1), 58-76. Retrieved from: http://digitalcommons.fiu.edu/hospitalityreview/vol27/iss1/4

Booyens, I., & Rogerson, C. (2016). Responsible tourism in the Western Cape, South Africa: An innovation perspective. *Turizam Medunarodni Znan, 64*, 385-396.

Burns, N., Paterson, K., & Watson, N. (2009). An inclusive outdoors? Disabled people's experiences of countryside leisure services. *Leisure Studies, 28*(4), 403-417.

Chesshyre, T. (2002). Independent spirit? This is for you. *The Times*, p. 4.

Chikuta, O., Forbes, K., & Chikanya, N. (2017). Tour guides experiences with tourists with disabilities. *European Journal of Tourism, Hospitality and Recreation, 8*(2), 130-139.

Efstathiou, A., & Zippelius, E. (2019). Cultural tourism, destination branding and spatial experience design: contemporary hospitality design in Cyprus. *Strategic Innovative Marketing and Tourism.* Cham.

European Regions Airline Association. (2018). *Passengers with Reduced Mobility (PRMs).* Retrieved October 23, 2021, from ERA (European Regions Airline Association): https://www.eraa.org/sites/default/files/180924_th_era_prm_brochure_online.pdf

Eurostat. (2014). *Glossary: Tourism.* Retrieved October 20, 2021, from Eurostat Statistics Explained: https://ec.europa.eu/eurostat/statistics-explained/index.php?title=Glossary:Tourism

Figueiredo, E., Eusébio, C., & Kastenholz, E. (2012). How diverse are tourists with disabilities? A pilot study on accessible leisure tourism experiences in Portugal. *International Journal of Tourism Research, 14*(6), 531-550.

François-Lecompte, A., & Prim-Allaz, I. (2009). The French and sustainable tourism: proposal for a typology. *Management & Avenir, 29*(9), 308-326.

Gillovic, B., & McIntosh, A. (2015). Stakeholder perspectives of the future of accessible tourism in New Zealand. *Journal of Tourism Futures, 1*(3), 223-239.

Goodwin, H. (n.d.). *Responsible Tourism.* Retrieved from Harold Goodwin: Taking responsibility for tourism: https://haroldgoodwin.info/responsible-tourism/

Goodwin, H., & Francis, J. (2003). Ethical and responsible tourism: Consumer trends in the UK. *Journal of Vacation Marketing, 9*(3), 271-284.

Holloway, C. J. (1989). *The Business of Tourism.* London: Pitman Publishing.

Kabii, F., & Wandaka, K. M. (2017). Effects of Tour Guides Training on Their Performance in Kenya. *Journal of Tourism and Hospitality Management, 5*(6), 233-250.

Kainthola, S., Tiwari, P., & Chowdhary, N. R. (2021). Overtourism to Zero Tourism: Changing Tourists' Perception of Crowding Post Covid-19. *Journal of Spatial and*

Organisational Dynamics, IX(2), pp. 115-137. Retrieved from https://www.jsod-cieo.net/journal/index.php/jsod/article/view/276

Katawandee, P., & Slisa, B. (2014). The use of ABC classification to analyse tourism receipts: A case of ASEAN tourists to Thailand. *Journal of Business and Behavioral Sciences*, 89-101.

Kim, S. S., Wong, K. K., & Cho, M. (2007). Assessing the economic value of a world heritage site and willingness-to-pay determinants: A case of Changdeok Palace. *Tourism Management, 28*(1), 317-322.

Krippendorf, J. (1984). *The Holiday Makers: Understanding the impact of leisure and travel.* Oxford: Butterworth Heinemann.

Langer, E. J. (1993). A mindful education. *Educational Psychologist, 28*, 43-50.

Loi, K. I., & Kong, W. H. (2015). *People with disabilities (PwD) in the tourism industry-Concepts and issues.* Institute of Tourism Studies. Macao: NYU Web Publishing. Retrieved October 28, 2021, from https://wp.nyu.edu/cts2015/wp-content/uploads/sites/657/2015/06/Loi-and-Kong.pdf

Mak, J. (2004). *Tourism and the Economy.* Honolulu: University of Hawaii Press.

Mindful. (2021). *Getting started with mindfulness*. Retrieved November 1, 2021, from Mindful: https://www.mindful.org/meditation/mindfulness-getting-started/

Ministry of Urban Development, Government of India. (2016). *Harmonised Guidelines.* Retrieved October 23, 2021, from Town and Country Planning Organisation: http://tcpo.gov.in/sites/default/files/TCPO/schemes/Harmonisedguidelinesdreleasedon23rdMarch2016.pdf

Moscarado, G. (1998). Interpretation and sustainable tourism: Functions, examples and principles. *Journal of Tourism Studies, 9*(1), 2-13.

Moscardo, G. (1996). Mindful visitors: Heritage and tourism. *Annals of Tourism Research, 23*(2), 376-397.

Muhanna, E. (2006). Sustainable tourism development and environmental management for developing countries. *Problems and Perspectives in Management, 4*(2), 14-30.

OHCHR. (2006). *Committee on the rights of persons with disabilities.* Retrieved October 24, 2021, from United Nations Human Rights Office of the High Commisioner: https://www.ohchr.org/EN/HRBodies/CRPD/Pages/ConventionRightsPersonsWithDisabilities.aspx#4

Poria, Y., Reichel, A., & Biran, A. (2006). Heritage site perceptions and motivations to visit. *Journal of Travel Research*(44), 318-326.

Pryor, C. (2001). Japanese pack for a longer stay. *The Australian*, p. 5.

Sandaruwani, J. R., & Gnanapala, W. A. (2017). Food wastage and its impacts on sustainable business operations: A study on Sri Lankan tourist hotels. *Procedia Food Science, 6*, 133-135.

Scutte, C. (1997). ASTA panelists offer pointers on booking independent travel. *Travel Weekly, 56*, p. 24.

Small, J., Darcy, S., & Packer, T. (2012). The embodied tourist experiences of people with vision impairment: Management implications beyond the visual gaze. *Tourism Management, 33*(4), 941-950.

Stanford, D. (2006). *Responsible tourism, responsible tourists: What makes a responsible tourist in New Zealand?* Wellington: Victoria University of Wellington.

Swarbrooke, J. (1999). *Sustainable Tourism Management.* Wallingford: CAB International.

Swarbrooke, J., & Homer, S. (1999). *Consumer Behaviour in Tourism.* Oxford: Butterworth-Heinemann.

Teo, C. B., Khan, N. R., & Rahim, F. H. (2014). Understanding cultural heritage visitor behavior: The case of Melaka as World Heritage City. In I. 2013 (Ed.), *Procedia - Social and Behavioral Sciences. 130*, pp. 1-10. Elsevier.

United Nations. (2010). *International Recommendations for Tourism Statistics 2008.* New York: United Nations. Retrieved November 15, 2012, from https://unstats.un.org/unsd/publication/seriesm/seriesm_83rev1e.pdf

Vasudevan, V., Vijayakumar, B., & Saroop Roy, B. (2017). *An Introduction to the Business of Tourism.* New Delhi: Sage Publications India Private Limited.

Végh, Z. (2005). A segítséggel élők turisztikai lehetőségei Magyarországon. [Tourism opportunities of people living with assistance in Hungary]. *Turizmus Bulletin, 4*, 26-38.

Weiler, B., & Ham, S. (2002). Tour guide training: A model for sustainable capacity building in developing countries. *Journal of Sustainable Tourism, 10*, 52-69.

World Health Organisation and World Bank. (2011). *World report on disability.* WHO. Retrieved from https://www.who.int/teams/noncommunicable-diseases/sensory-functions-disability-and-rehabilitation/world-report-on-disability

Yau, M. K., McKercher, B., & Packer, T. L. (2004). Traveling with a disability: More than an access issue. *Annals of Tourism Research, 31*(4), 946-960.

Zhang, H. Q., & Chow, I. (2004). Application of importance-performance model in tourguides' performance: evidence from mainland Chinese outbound visitors in Hong Kong. *Tourism Management, 25*, 81-91.

Zsarnóczky, M. (2018). The future challenge of accessible tourism in the European Union. *Vadyba Journal of Management, 33*(2), 39-43.

2

3 Communication for Tour Guides

Learning objectives

After studying this chapter, you should be able to understand:

- The process of communication with a tour guide as a communicator.
- How tour leaders and tour guides should present themselves
- Cross-cultural differences and how to handle them

This chapter focuses on the preparations that an individual needs to make to become a tour guide. A tour guide or leader must hone their communication skills at a conceptual level. Specific attributes of the tour guide will make the communication effective. We will discuss how professional tour leaders and guides should present themselves to appear credible. It is vital that the communication is contextually relevant. Often guests are from a different cultural background. For communication to be effective, tour guides must be sensitive to the cultural make-up of the guests and factor the same in their communication.

Communication

Tour guiding is essentially about communication with the visitors to help them negotiate the place with understanding and education. Effective communication is, therefore, the foundation of effective interpretation. A tour guide or a tour leader will often interpret gaze in different situations – a historical site, a walking tour, a church, a museum, or a natural setting. However, the basic principles of communication will remain the same, and a tour guide or leader must perfect their interpersonal communication skills.

According to the dictionary, communication is a process by which information is exchanged between individuals through a common system of symbols, signs, or behaviour (Merriam-Webster, n.d.). In the case of a visitor, the purpose of sharing information is to associate meaning to what is being seen or experienced – a gaze, an entity, an activity through a language, verbal and non-verbal, that is mutually understood. Interpretation is, therefore, about creating and communicating meaning.

Communication has two ends – the sender and the receiver. In a touristic interpretation scenario, a tour guide is often the sender while the visitor is the receiver. The visitors must receive in totality what is communicated by a tour guide. Often some distractions and barriers distort the communication. The effectiveness of communication also depends on the receiver's needs. Multiple needs related to the visit exist at any time, and they need to be addressed by the interpreters.

Maslow's hierarchy of needs

Abraham Maslow was among the early psychologists to explain human motivation with his theory of a hierarchy of needs (Maslow, 1943). His premise was that there are different types of human needs, hierarchically ordered. Needs are a source of motivation, and arise when the need at the previous level is satisfied. This argument had no scientific basis and was refuted by other scholars, but Maslow's need-hierarchy helps us point towards a hierarchy of visitor needs that the tour guides and tour leaders must take care of.

- According to Maslow, at the lowest rung of the pyramid are the basic or physiological needs – food, clothes and shelter. Basic needs are the primary physical requirement for human survival and prelude any other higher-order need.
- The next rung is the safety and security needs. Once the physiological needs are primarily addressed, safety and security needs become dominant. An individual is motivated to satiate these needs.
- The third rung is social belongingness, or simply social, needs. These include affection, acceptance, and belonging. At this level, the need for interpersonal and emotional relationships are the driver.
- At the fourth level of this need-hierarchy is esteem. Once the physiological, safety and social needs are addressed, an individual looks for respect and appreciation by others. They want their accomplishments to be recognised and direct their efforts accordingly.
- At the top of the pyramid is the self-actualisation needs. These include self-fulfilment, personal growth and realisation of one's potential. Maslow later clarified that the needs are not mutually exclusive. When a need in a deficit is more or less addressed, it ceases to be a motivator, and the individual is likely to be motivated by achieving the next set of needs.

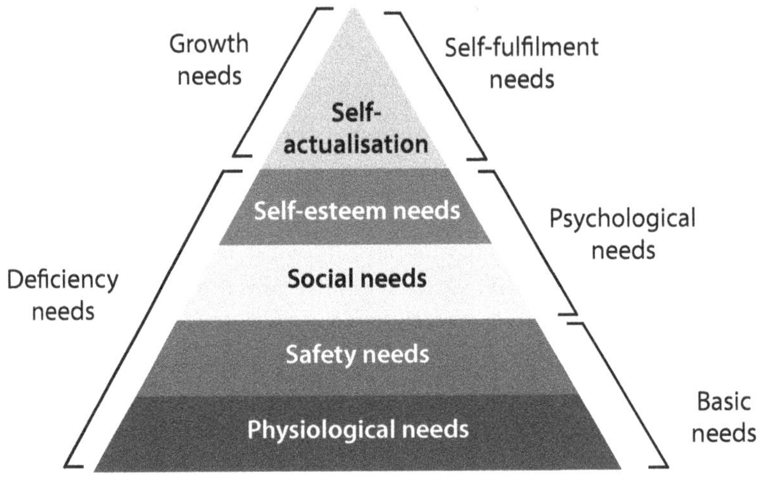

Figure 3.1: Maslow's hierarchy of needs

Maslow's theory of need hierarchy has been subject to many critiques. However, the discussion on this theory leaves us with some pointers. One, the entire hierarchy can be broadly divided into three essential types: basic needs (physiological and safety), psychological needs (social and self-esteem) and self-fulfilment needs (self-actualisation). Tour guides will have patrons whom these needs would drive – some more concerned with their safety and security or physiological well-being; some would be more concerned with being with their group or merely the Instagramability of their visit; others will be more concerned with their visit learning and knowing.

Two, the drivers of the visit to a place can be divided into deficiency needs (basic and psychological) and growth needs (self-fulfilment). Only when the deficiency needs are addressed, the visitor will derive value from the growth needs. This means if the basic needs are not addressed, as if the guests are hungry, tired or feel unsafe, they will not pay attention to the tour guide's commentary or brief. Many visitors are interested in clicking photos and posting them on social media. An itinerary should be so planned that there is sufficient time for photos and selfies. The tour guides must be proficient in photography and even advise the guest about the best view and angle. Overdoing these needs might not help, but a deficiency will undoubtedly hurt.

On the other hand, visitors are interested in knowing about the place. They would like to immerse in local experience to understand it better – in tourism parlance they say "live like a local". This phenomenon is self-actualisation. This learning, an essential component of travel, is a growth need. However, an individual typically pursues this growth need only when the 'deficiency needs' are no longer deficient.

The third critique is that different needs can co-exist. For different individuals, even within the group, the gravity of different needs is different, and many needs can co-exist at any given point in time. However, since these needs are a common concern, a tour guide must ensure that deficiency needs are addressed before attending to the growth needs.

For example, a guest standing in the sun for long would feel uneasy and not focus on what the tour guide says. Similarly, a tired or hungry guest may not listen with interest. The first thing visitors may like to know at a site is restrooms, cafes, etc. A tour guide may brief about this. The next concern is about safety and security. A guide must ensure that the guests are safe – especially when outdoors – during a city or wilderness walk. They must brief the guests about safety dos and don'ts and ensure that the safety protocols are adhered to. They must explain the traffic rules or tips to cross the street. Similarly, the guests must be advised about wild animals or rugged terrain in the wilderness. A carefree guest is more likely to engage with the tour guides. Participation and group dynamics are all concerns of a professional tour guide. Later in the book, we will see how visitors' needs must be addressed.

Basic communication processes

With an eye on the guests' varying needs, a tour guide must communicate effectively. Communication follows a typical process. Tour guides can ensure their messages' efficacy by managing the communication process – comprehension, understanding, and retention. It is a continuous and dynamic process affecting and being affected by many factors. The most accepted communication process model includes sender, receiver, encoding, decoding, message and feedback. Then there are various channels or mediums of communication like audio, video, electronic channels, written or body language.

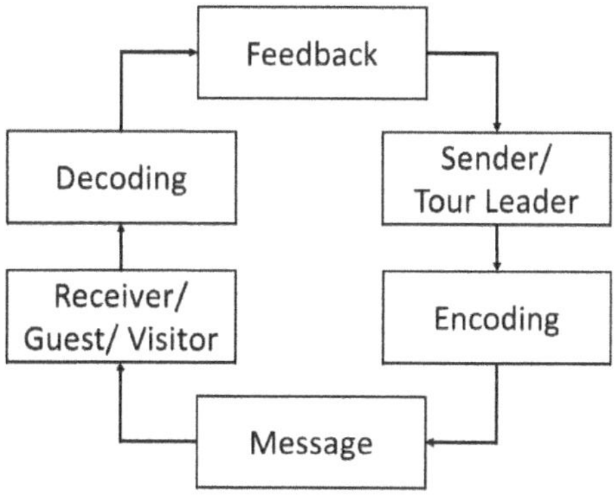

Figure3.2: The basic communication process

Let us discuss each of these components.

Tour guide – the sender

A sender initiates the communication and sends a message to the receiver. In our case, the tour guide or tour leader is the message's sender. One of the primary issues in this scenario is why a tourist should trust a tour guide. The answer is simple – because the tour guide is trustworthy or credible. If the sender is credible, the likelihood of the message being received well and accepted is greater. The perceived credibility of the guide is the most critical characteristic. For example, a shabbily dressed tour guide or one who fails to look into the audience's eyes, though an expert, would fail to convey credence. Credibility is created through three elements: content, confidence, and commitment.

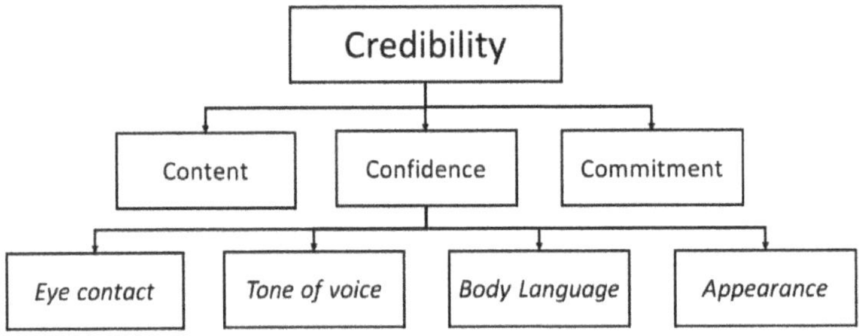

Figure 3.3: Credibility of sender

Content

The basic expectation of a guest is that the tour guides present credible information. A guide is assumed to be an expert of the site and is expected to be accurate with the facts. Much information about the site will be readily available to the visitors, and they hire a tour guide so that they get the relevant information interestingly delivered to them. If a guide is caught unaware, it will severely dent their credibility. Often, tour guides twist facts or present folklore as facts to make commentary entertaining. Any misrepresentation or distortion in the presentation of facts would be readily figured out and is considered a cardinal sin. A tour guide should thoroughly research the content before presenting it. However, a guide cannot know everything. It is Okay to convey that you do not know something and will soon return with more information.

Examples of credibility expectations from tour guides

Excerpts from *Professional Ethics for Tour Guides in Hong Kong*

III.4.iii. Be knowledgeable about Hongkong.

III.4.iv. Provide visitors with accurate information

(TIC, 2021)

> Excerpts from *Tour Guide Code of Conduct* (Travel Wild)
>
> Customer Interaction
>
> I will make every effort to present true and accurate facts and ensuring that a clear distinction is made between stories, legends, traditions and opinions.
>
> (TravelWild, 2017)

Confidence

A confident tour guide is considered credible. Eye contact with the guest, body language, tone of voice and appearance are ways to appear confident.

Eye-contact

When people look into each other's eyes, it is called eye contact. In social psychology, eye contact is considered a non-verbal communication that displays confidence, respect, and social communication (Dictionary.Com, 2021). Studies have endorsed that people who make eye contact are perceived as more honest, warmer, more knowledgeable (Brownell, 1982) and more confident (Ruth, 2017). Tour guides are advised to establish eye contact with their audiences as they speak. This makes them look confident and, therefore, credible. Though the concept of eye contact is construed differently in different cultural settings, of late, the western interpretation of eye contact is generally accepted. Not making eye contact may be interpreted as suspicious or not having the courage to look at the other.

However, eye contact does not mean a continuous stare at the guests, which would be rather embarrassing. It is advised to look into the listener's eyes for two to three seconds, focus on the next person and then someone else, and return to the earlier ones. If it is a large group, do not look at only one part or side. Try to span and reach the entire group. Talking to one person or only a group section will create perceived inequity, and others may feel left out.

Body language

Though often misquoted and misinterpreted, Albert Mehrabian's 7% – 38% – 55% rule is often cited in human communication (Mehrabian, 1981). It is taken as meaning that people deduce our feelings, attitudes, and beliefs, 7% from spoken words, 38% from the tone of voice and 55% from the body language. Thus, the tone of voice and body language are essential to human communication. Body language is a non-verbal form of communication where physical behaviours such as postures, gestures, facial expressions, eye movements, touch, and space convey a message. Many researchers believe that body language is more communicative than verbal communication. Up to 60-90% of meaning is conveyed through body language (Jurin, Danter, & Roush, 2000). Brownell (1982) says that this form of communication is more accurate and authentic.

While most people use body language subconsciously, the tour guides should practice and communicate appropriate body language to convey their message and suggest that they control the situation. Tour guides should maintain an assertive posture. They must generally stand straight with feet in an open and wide stance. The chin and head should be up and firm. Hands do much talking, and professional tour guides should learn to know what their hands are saying (Pachter, 2013). They should not keep their hands in pockets or speak with hands folded. Hands should generally be visible. They must remain conscious that they are not fidgeting. Tour guides should appear interested (in the visitor) rather than making efforts to sound interesting.

For instance, a tour guide who frequently looks at his watch and folds his hand across his chest will not connect with the guests.

Tone of voice

The third dimension of confidence is the tone of voice. This is the pitch or modulation of voice, that is, the rise and fall of the sound. Put simply, tone of voice is how one speaks. Excitement and feeling in the tone of a voice create associated meanings in the message. Therefore, tour guides must be sure that the emotion in their tone of voice matches the intended message.

There are two types of vocal tonality – dominant and submissive. Tour guides must have dominant tonality in their commentaries which exudes leadership, assertiveness and security. Visitors are more likely to trust and respect the tour guide. They must avoid vocal 'fry' and 'speak-up', which are considered symptoms of low credibility. Vocal fry is a way of speaking where the voice is low pitched and has a characteristic rough or creaking sound. Speak-up is using a higher pitch at the end of the sentence. Similarly, speaking very quickly without pauses reflects nervousness. Speaking too softly or too loudly is also considered as low confidence. Avoid frequent use of 'ums' and 'uhs', which is also considered not good.

There is no one best way of speaking. Tour guides should make their commentaries enjoyable by voice modulations and dramatic pauses. In general, the voice should be audible, free-flowing and moderately paced so that a non-native listener can also follow what is being said. Style of speaking is a nervous system habit formed over time and under the influence of many – parents, teachers, seniors, and people one idolises. It is also about practice and conscious efforts. Receiving feedback and improving the tone of voice is helpful.

The tone reveals a lot about the tour guide's attitude – whether they are interested in guiding, whether they are tired, confident, or distracted. Often, we feel that the tourists do not notice this. However, that is not the case.

Appearance

The physical appearance of the tour guides influences their perception of credibility. Hamermesh (2011) suggests that good-looking people are more effective than others. Physical appearance is a predictor of self-esteem. People with high self-esteem are conscious of their physical appearance. It is about how much they like themselves and their attributes and qualities. Those with high self-esteem exude positiveness and confidence and tend to be more sociable, and this is what a good tour guide should be like.

On the other hand, people want to communicate with those who look good. Therefore, tour guides should present themselves as confident professionals by wearing comfortable and presentable clothes. Wearing a tidy uniform communicates authority and expertise.

Commitment

The third ingredient of credibility is the commitment reflected in the passion and sincerity with which a tour guide talks with his guests. Use of common words and straightforward language are indicators of sincerity. When tour guides are sincere, they talk straightforward, and thus communicate effectively. Forthrightness cultivates mutual respect between the tour guide and the guests leading to trust, making the guests more receptible. Guests tend to forgive many slip-ups and mistakes if they believe that the tour guide is sincere and passionate. Tilden (1967) also identified passion and sincerity as essential attributes of an accomplished tour guide.

Valuable tips for creating credibility

- Build personal relationships through the exchange of non-business pleasantries.
- Ask questions, but do not put pressure
- Avoid using jargon
- Summarise and test understanding
- Use verbal and non-verbal techniques to support ideas, e.g. use `I'language to convey that the speaker accepts responsibility for what he or she is saying.
- Maintain eye contact with others as eye contact communicates sincerity, concern and involvement
- Show a relaxed attitude towards time. Do not place unnecessary pressure on guests and self by creating artificial deadlines that cannot be met. Tourists are here for leisure.
- Use your face to show pleasant feelings, smile or mild laugh
- Dress elegantly; be neat and presentable
- Demonstrate personal hygiene; be clean and have no offensive odour.

Moved from cultural differences section

Commentary – the message

The message refers to information being given to the receiver by the sender. Often the tour guide would like to inform the guests about things or events in the gaze. They may have to brief the guests, instruct them, or even handhold them through an experience. In a tourism situation, communication must be within a context or a servicescape where other non-personal communication elements like displays, descriptions, notice boards, signage, and others must be accommodated in the messaging.

Encoding

Before transmitting, the message must be encoded. Encoding also includes placing the message in an appropriate medium – verbally or non-verbally. The context – time, space, situation, and the nature of the message is also important. Thus, depending on the context, the message is organised as a combination of words, symbols and images that can be effectively communicated. For example, the tour guide may use symbols or signs in a windy outdoor where the sound is not carrying effectively. Similarly, in a place of worship where the tour guides may not be comfortable speaking, they may point out a painting or a description provided.

One of the crucial concerns is that the visitors are often from a different linguist background. This situation requires linguist tour guides and leaders to speak the guests' language and have cultural sensitivities. Therefore, a message coded in the receivers' language and within their cultural context will be compelling. We discuss this in detail in the following section.

Inappropriate encoding may defeat the entire intent of the communication.

Visitors – the receivers of the message

In our case, the individual guest or the group, who might be a listener, reader, or viewer is the message's intended receiver. If the receivers are not appropriately situated, they may not receive the message correctly. There may be external distractions like noise or some commotion. The receivers may be tired, hungry or not in a good mood. Sometimes they are not able to follow the leader because of language or an accent barrier. The guests' comprehension of the message depends on their prior understanding of the subject matter, relationship, and trust with the tour guide. Thus, it is incumbent upon the tour guides to develop trust and camaraderie with their clients as soon as possible. The guests might take some time before they start following the tour guides' communicative style in general and accent in particular. Later in the book, we discuss methods followed by the tour guides to be more communicable.

Receivers of the message – the guests – often visit a place to learn about it. Learning is about deciding to know. It is a two-step process: first gathering information and then making sense of it. Thus, two issues are relevant. How does one gather information? How does one process that information?

There are two ways of noticing data. First, one is absorbing it literally through five senses – hearing, touch, sight, taste, and smell. Examples may include seeing a sculpture, hearing the roar of lions in a wildlife sanctuary, touching a coral, tasting the local food or smelling a flower in a garden. The other way of capturing data is intuition, which generates abstract possibilities from information gathered, like drawing meaning from what we see locals doing, or imagining while visiting a historical site how people lived there. Most people use a combination of sensing and intuition to different degrees of effectiveness while obtaining information at any point in time. Then, based on the processed information, one tries to make sense of it. There are two ways of doign this: thinking and feeling. Thinking is arriving at a conclusion based on reason and logic. However, the conclusion can also be drawn based on one's value system or beliefs – this is feeling. Again, an individual often uses a combination of thinking and feeling to arrive at conclusions based on how they have accessed the information – sensing or intuition. Often guides must advise visitors to follow established protocols for performing specific tasks or advising them about right or wrong. This approach is the thinking approach. However, tourists engage in shopping based on their feelings. They may also visit some places because of a historical or emotional attachment.

As early as 1913, in his masterpiece, *Psychological Types*, Carl Jung discussed the four personality types that bear an individual's communication style. A quick assessment of the kind of receiver/s that tour guides have among their audiences will help them adapt their communication accordingly. His four personality types include thinkers, feelers, intuitors, and sensors (Jung, 1913).

The *thinker* type analyses consciously and deduce information. They are comfortable with the use of data that includes facts and figures. Thus, these guests would gain more understanding when the tour guides organise information logically using facts and figures to make their point. On their part, thinkers will evaluate what the guide is saying, whether that sounds logical, the source of data, and how that relates to them or the information they have about their situations. For this type, the guide may frequently refer to the sources of his information. Thus, one can identify thinker-guests as those who continuously seek logic behind everything being said. They also contribute by bringing in facts and figures. At times they will point out inaccuracies in what the guide says and want them to be acknowledged and rectified. Thinkers may take some time to absorb what is being said. Thus, thinkers are analytical and organised.

Analytical thinkers may allow tour guides time if they present logically organised details step-by-step. However, vague or cooked-up interpretations put off these proof-seeking sceptical guests. Tour guides should avoid exaggeration, informality, and attempt to hard-sell an opinion or a particular point of view. They are impressed with facts and references.

Feeler types process information and draw conclusions based on their emotions. They have a sense of 'feeling their way'. While feelings are the conscious consequence of emotions, emotions are unconscious. Feeler types will typically work to internalise the 'feel'. For example, they would try to connect with how they enjoyed similar places and situations earlier or what they had planned they would do in this situation – this can take some time. Feelers prefer a personal communication style, and like to listen to a version of the place or situation from a guide who represents the local community. One can recognise feeler-guests as those who are friendly and empathise. They are generous in using words to express emotions. The guide needs to be patient with this type and allow them time to internalise the place. Thus, feelers are considerate and friendly.

Friendly feelers allow guides enough time without haste. They prioritise an amicable relationship both with the guide and the place. They look forward to relaxed slow-paced interaction. Exchange of pleasantries is appreciated, and commentaries should not be too fast-paced. Attention-seeking feelers love guides who take an interest in their lives.

The *intuitor* types do not deduce a conclusion logically, but draw intuitive conclusions. These guests do not know how they arrived at a conclusion yet believe they made a correct decision. A sense of automatically 'knowing the answer' is the conscious consequence of intuition, an unconscious process. For example, an intuitor guest might not be very interested in a lengthy commentary by a tour guide. They believe they know sufficiently about the place and are here to see it rather hurriedly; thus, they prefer an overview rather than a lengthy flat commentary. Intuitorss too can be recognised. They are rather conceptual, and it is a challenge for others to understand them because they talk with little reference to the context, assuming that everyone knows what they are talking about. If someone asks a question, they get irritated, not realising that others do not know what is in their mind. So, a tour guide should begin with a brief overview. Then, if these guests demonstrate any interest, a detailed explanation could follow. Thus, intuitors are creative and expressive individuals.

Expressive intuitors also appear to be short of time, not seeking too many details (they believe they already know). They look forward to brisk and stimulating interaction with the tour guide, who should avoid any solemn canned detailed presentations. Do not talk of too many rules or too many facts. They do not like to be interrupted when talking (actually when others want the guide to be talking).

Often intuition is referred to as a gut feeling, presenting it as irrational. However, intuition is a somewhat rational process. It is like thinking but is a subconscious phenomenon. Intuition being subconscious, people are not aware of this activity, which does not mean it is irrational. Instead, people subconsciously and fragmentedly process information from different sources to create their impression of the subject matter.

The *sensors* type use the input from their senses to process information and conclude. They would like to see the whole picture and may take some time doing so. Drawing conclusions is thus a conscious process. This type would somewhat be interested in knowing and participating but would also keenly observe what others are doing or what other tour guides are advising their group and might like to change their agenda accordingly. Therefore, it is an excellent strategy to provide them with some brief basic information such that it helps them have an idea of the complete context. The sensor type guests will use this for orienting themselves and will not have to look around. One can identify sensor guests. They are conscious of time consumed and often convey the concern for time. They want the itinerary to be completed on time and might appear in a hurry and tend to suggest that you, the guide, are taking too much time in niceties or too many details. They have a short span of attention. Deadline driven sensors are thus decisive and doers.

Sensors are forceful, direct, poor listeners who look for quick, to the point, direct talk with few details. Therefore, guides should avoid being overly friendly. Cross-selling or oversharing should be avoided in all cases.

For analysis and discussion, we use Carl Jung's classification. However, at different times an individual is a combination of all types – some thinking, some intuition, some feeling and some sensing.

Decoding

The receiver, having received the message, decodes it. This is the heart of interpretation. Communication is successful only if the receiver decodes the message as it was intended to be. The communication will fail if the receiver does not have the tools to decode the message. The sender must therefore be sensitive to the decoding abilities of the receiver. As a simple example, no matter how good a French-speaking tour guide is, she will be ineffective if her Chinese or Indian group does not understand French. Generally, over time, people develop the ability to decode various messages. Different people decode messages based on their judgments, reminiscences, and mechanisms, creating their own meanings. A professional tour guide should remain aware of their audiences' abilities to decode. They must encode messages accordingly.

Feedback

One of the essential components of the process of communication is feedback. The sender must seek feedback while the receiver must provide it to ensure that the message is as close as possible to what the sender intended. The more the two parties use feedback, the more effective communication will soon become.

Communication is not a one-off activity. When an idea is expressed, it creates ripples through time. It sets the stage for the next bit of communication. Communication needs to keep moving, as the receiver becomes the sender of the feedback by responding and reacting. This feedback allows the sender to compliment the original message and shape the subsequent communication.

The tour guide needs to keep the group together by regularly asking guests if they have been following the information, and trying to watch the person who seems to be wandering around the most. These tourists are the ones who have difficulty focusing and later report dissatisfaction with the tour guides.

Handling questions

Handling questions is an essential technique for tour guides. Whenever a customer poses a question to a guide, the guide should not respond immediately and reply directly the questioner. In these situations, a couple of things come into play. First, guides must not talk to individuals. Instead, they must talk to the group. If the guide responds directly to questioners, they tend to engage with the guests who ask frequent questions and others may feel left out. So, they must transfer the query to the entire group.

John (to the guide): Was the red fort in Delhi and Agra built by the same Emperor?

Tour guide (to John): Thanks, John, for that good question.

Tour guide (now spanning the entire group): John wants to know if the two red forts, one in Delhi and the other here in Agra, were built by the same Emperor.

Tour guide (looking at John): Was that the question, John?

(Tour guide explains)

(After the answer)

Tour guide (looking at John): Have I answered you?

John: (Nods in affirmative)

Thus, one, the tour guide transfers the question asked by John to the entire group and engages their attention. Two, the guide appreciates John for the question encouraging John and others to ask more questions because that helps take the talk further and in a direction that guests want. Third, the guide paraphrases the question and takes feedback from John to be sure that is the correct query. Finally, the guide again takes feedback to confirm if the query is addressed after the explanation.

Medium and noise

One of the crucial concerns in communication is medium and noise. Medium refers to the channel chosen for communication – verbal or nonverbal; it could well be an e-channel. The sender must choose an appropriate medium for communication. In a typical tourism situation for the tour guide, communication is mostly a combination of verbal and nonverbal modes.

Noise refers to the impediments, sender and receiver face during the communication process. These barriers can mean real physical noise, poor comprehension of the message due to predisposition of the sender or the receiver or their body language, inattentive receiver, and other impediments like language and cognitive precision.

While the guest's abilities would be a limitation, to eliminate or reduce noise, the tour guides must factor in the situation and sources of noise before configuring the message to be communicated through the most appropriate medium.

Understanding cultural differences

Every day is a new day. Every tourist is a different person. Meeting unknown persons and serving them would always be a challenge. Ideally, tour guides would like to know as much as possible about their guests beforehand – their tastes and preferences – so that they can organise themselves to serve them better. However, this does not happen. Every tourist or group of tourists is different from the previous one. They come from different countries, different cultures, different religious following, different socio-economic backgrounds and each with their peculiarities. These differences shape their expectations of a holiday and their expectations from a tour guide. As a professional tour guide, one needs to live up to guest expectations.

Always acquaint yourself with the group. Ask questions if you must to know about their background. For tour guides to address an individual's needs, it is essential to understand the differences in their background. It is, therefore, necessary to understand the primary ways in which cultures around the world vary. However, guides have very limited time to do so, and they must be equipped with an understanding of cultural differences beforehand.

The bad news: Hundreds of cultural and sub-cultural constructs are difficult to learn and shape your demeanour. Further, people would be different even from within a broader cultural group. For example, not all Indians are alike. Neither are Britons or Germans or Russians or Latin Americans or all Africans.

The good news: Tourists guides should take note of their destination. Every destination is popular with particular source markets; over time, a destination becomes incredibly popular for tourists from selected countries and cultural backgrounds. For example, some places attract more Swedes and Finns while others attract Russians or Germans. In India, some destinations are more populated by tourists from Gujarat, while others are frequented on weekends by professionals from Delhi. Based on the tourist profile of your destination, you may initially try to understand the primary ways in which your targeted customer groups behave and conduct themselves.

Culture has many definitions. While some look at it as an integrated system of learned behaviour patterns that distinguish characteristics of the members of a given society, others understand it as collective programming of the mind which distinguishes the members of one group from those of another. From our perspective, culture shapes behaviour or structures one's perceptions of the world. This approach suggests two things: one, tourists from different cultural backgrounds will have different perceptions and will decipher host culture and heritage differently, and as tour guides, we need to understand their cultural context to connect well to them; two, we also need to conduct ourselves in such a manner that our behaviour does not offend them or is irritating for them.

Unfortunately, most people consider the cultural difference as a difference in observables – anecdotes, arts, ceremonies, communications, heroes, habits, jargon, language, myths, norms, physical arrangements, rituals, dresses, crafts, food, and many others. These tangible differences are just the tip of the iceberg. Culture is much more deep-rooted, and is about shared values and beliefs. These are not directly observable but can be distilled from how people explain what they do – these include cognitive process, beliefs, commitment, ethics, feelings, values,, vision, knowledge and sentiments. Values and beliefs do not exist in isolation. They are grounded in the philosophy of life shared by the group – assumptions, consensus, ideologies, mindset, world view, among others.

Cultural differences are differences in fundamental ways in how people think and what people believe. For example, the Punjabi community from north India is known for its exuberant lifestyle. They are fun-loving people who appear to enjoy the present rather than care for the future. They are spend-thrift and pompous. Here is an exciting explanation. One Punjabi friend once quoted a couplet that captures the essence and philosophy behind this `*Khaya-piya lahe da, baki ahmed shah-e-da* '. Which means 'whatever you consume (eat and be merry) is yours; everything else will go to Ahmed Shah'. Ahmed Shah Durrani, aka Ahmed Shah Abdali (1722-1772), the founder of the Durrani empire of Afghanistan, plundered India nine times, including the 'great killing' on February 5, 1762. Every time he came to India and returned, he passed by Punjab plundering and looting the people. Thus, this couplet advises material consumption and to live in the present for else everything could be lost. This understanding over time has become the life philosophy of the people in the region. They might have forgotten the history, but the enshrined values and beliefs have passed from one generation to another.

Tour guides, therefore, must first try to know who their customers are. Where do they come from? However, never forget that individuals in a culture can be diverse, and stereotypes can describe not all individuals. Understanding the main characteristics of predominant cultures of this world can help us inform and educate ourselves on how better to facilitate communication with guests from different cultural backgrounds.

Survival tip

Tour guides need not sit in judgment over the cultural differences. `What we do is right (because I have always done that!) and others are wrong'. This kind of belief recoils on oneself to create anxiety and disillusionment, manifest in inappropriate behaviour. The key is to understand that it is not a question of right or wrong but difference. That is it! Appreciating the difference does much good to you – it renders you comfortable in the company of others from a different culture, allowing you to learn from these cross-cultural interactions, and it makes you more open to understand the expectations of the tourists and adjust accordingly to make them comfortable in your company.

Appreciating the cultural differences

A tour guide must be aware of and sensitised to the differences a traveller feels. The four primary ways a visitor confronts an alien culture are time, space, context, and religious beliefs. Time is understood in terms of monochromic vs polychromic cultures. We will talk about space in terms of *proxemics* (use of space and proximity), *haptics* (touch), *kinesics* (body movement) and *vocalic* (paralanguage). We will analyse high context and low context cultures to interpret the context. Finally, it is also essential to understand the religious differences.

Monochronic vs. polychronic cultures

Chronemics is the study of how time is used for non-verbal communication. Chronemics suggests that a vital communication approach is how we perceive time, structure our time, and react to time. Time perceptions can be expressed through punctuality, willingness to wait, speed of speech, or even the amount of time people are willing to listen (buzzle.com, 2013). For example, arriving at a party on time is impolite in Argentina. So instead, arrive 30 to 60 minutes late.

Cultures can be put into two-time system categories – *monochromic* and *polychromic*. A monochronic time system means that things are done one at a time, and time is segmented into precise, small units. This interpretation of time is scheduled, arranged, and managed. Monochronic cultures typically include Germany, England, Turkey, East Asia, Japan, Jamaica, Canada, Switzerland, the United States and Scandinavia.

A polychronic time interpretation is an approach where several things can be done simultaneously, and a more adaptable attitude towards scheduling time is needed. Polychronic cultures include Saudi Arabia, Egypt, Mexico, New Orleans, the Philippines, Pakistan, India, and Africa.

For monochronic cultures, `time is tangible' and viewed as a commodity where 'time is money' or 'time is wasted'. The result of this perspective is that monochronic cultures place a paramount value on schedules, tasks and 'getting the job done' – so should the itinerary. These cultures are committed to regimented schedules and may view those who do not subscribe to the same perception of time as disrespectful (Guerrero et al., 1999). Therefore, punctuality is an absolute must for tour guides when the guests are from these backgrounds.

Polychronic cultures are much less focused on the preciseness of accounting for each and every moment. In polychronic cultures, time is elastic, and people have all the time in the world. Relationships and tradition are more important than time. A comparison of monochronic and polychronic people is given below, and implications for tour guides are as below:

Monochronic culture	Polychronic culture
Examples: US, England, Scandinavia, Canada, Switzerland, Germany, and Japan.	Examples: Latin American, African, Asian and Arabic cultures
Do one thing at a time	May do many things at once
Concentrate on the job	People are highly distractible and subject to interruptions
Take time deadlines & schedules seriously	Consider an objective to be achieved, if possible
Are low-context and need information	Are high-context and already have information
Committed to the job	Are committed to people and relationships
Adhere religiously to plans	Change plans often and easily
Concerned about not disturbing others; follow rules of privacy and consideration	Are more concerned with those who are closely related than with privacy
Show great respect for private property; seldom borrow or lend	Borrow and lend things often and easily
Emphasise promptness	Base promptness on the relationship
Are accustomed to short-term relationships	Strong tendency to build lifetime relationships

Implication for tour guides dealing with monochronic guests

This assumes that guides would handle relatively effortlessly, guests from polychronic cultures.

- Monochronic tourists will expect undivided attention of the tour guide and that he/ she does not involve in other things/ with other people
- They would like a tour guide to carry out the job with professionalism. Attending phone calls in between could be considered highly offensive.
- Punctuality is an absolute must. A tour guide should reach the meeting point at least 15 minutes in advance, and they should complete the itinerary as scheduled – not early, not late. Incomplete narrations will not be looked upon favourably.
- Do not assume that the tourist has the information. They would expect ample structured information.
- Do not entertain friends and acquaintances while you are with a group. This act is considered a non-serious, unprofessional attitude.
- Do not change itinerary and schedules without valid reasons and prior information
- Guides sometimes have an easy-going attitude. They prompt visitors to take shortcuts and bend the rules. This could be considered unprofessional rather than 'friendly'.
- Some tour guides unintentionally display lack of respect towards monuments, etc., like touching the walls/ painting or standing with support by putting a foot on the wall. Guests sometimes feel annoyed with this.
- Guest expects a prompt response to their problems. They expect someone to be on the job when they register a complaint, such as fixing a broken window on the coach.
- Be professional. Do not pile on a guest. They might be friendly with you, but that does not mean you expect a long term benefit or association from them.

High context versus low context cultures

One way to make sense of cultural differences is to understand them as high versus low context. Context can be understood as the stimuli, environment or ambience surrounding the environment (Hall, 2002). How a culture relies on the three points to communicate their meaning will place them in either high or low context cultures. Low-context cultures assume that the individuals know very little about what they are being told and must be given a lot of background information. High-context cultures assume the individual is knowledgeable about the subject and has to be given very little background information.

While individuals must be monitored and observed, broadly, nationalities can be assumed to fall in categories as below:

Low context cultures	*High context cultures*
American (northern states), Australian, English Canadian, English, German, Irish, New Zealand, Scandinavians	Africans, Arabs, Latin Americans, Southern Europeans (Italian, Spanish, French, Portuguese, Greeks), Southeast and Far east Asians, Turks and Hungarians, Russians, Finns, southern and mid-western US, French Canadians

Handling diversity

High context cultures, rather than low context cultures, display a need for higher in-groups and group reliance. The co-existence of subgroups is acceptable in high context cultures, whereas a group is the individual's identity. The small group relies on their common background to explain the situation rather than words. On the other hand, low context accommodates individuals with a wide variety of backgrounds. So, their cultural group identity is essential for high context nationalities, while for low context nationalities, individual identity is emphasised.

Tradition and history (time-orientation)

Higher context cultures tend to value tradition and history more than lower context cultures. Your narrations should therefore vary accordingly. On the other hand, for guests from lower context cultures, recent shared experience is the reality; therefore, they need to be explained with rationale and logic.

Dependence and relationships

An individual from a lower context culture seeks more autonomy than those from a high context culture. He/she may seek a relationship, but short term and intimate in only selected cases. High context individuals may consider them 'not reliable', or may be frustrated by people appearing not to want to develop a relationship or continue to help them on an ongoing basis.

The implication is that a dependent high context culture individual may ask more questions rather than venturing out independently. On the other hand, a low context culture individual may feel thwarted by handholding a tour guide from a high context cultural background.

Use of space and non-verbal communication

Cultural differences are also manifest in non-verbal communication, including the use of space. Failure to appreciate the difference may inadvertently be misunderstood. Non-verbal communication is a crucial component of interaction as it communicates a lot. However, it is more challenging to master this form of communication than oral and written communication.

Proxemics

Proxemics is about the use of space, and distances between people. The term was coined by Edward Hall who introduced the concept of personal territory.

This is the space that we carry with us, the `bubble' of space that one keeps between self and the persons around them. The personal territory has four zones or areas:

- **Intimate space**: ranges out to around one foot and involves a high probability of touching. We reserve it for whispering and hugging.
- **Personal space**: ranges from 2 to 4 feet and is used for interactions among friends and family members.
- **Social space**: ranges from 4 to 10 feet and is used for communication among business associates and other acquaintances.
- **Public space**: ranges from 12 to 25 feet and is the distance maintained between the audience and a public speaker where one can hear but is not tied to the speaker.

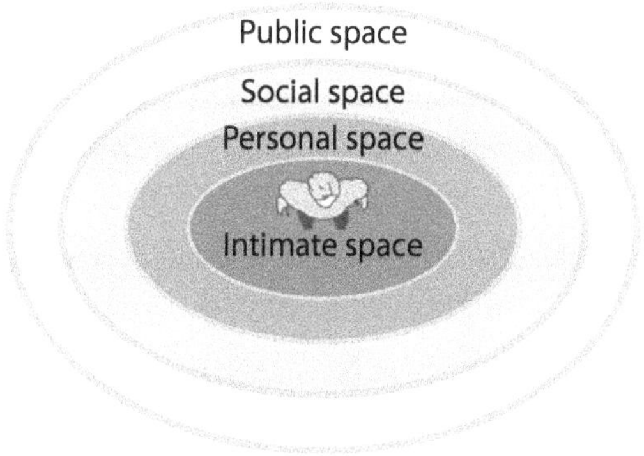

Figure 3.4: Hall's concept of interpersonal distances

However, personal territories vary both culturally and ethnically. Therefore, differences need to be appreciated. For example, male Saudi Arabian guests will almost come up to you as you explain a monument because their social space almost equals our intimate space. Sub-consciously, you would appear running away from them to maintain a decent distance in an attempt to regain your social space while your guest would invade you to regain his social space. Finally, you may interpret him as being 'too pushy' while he would consider you 'not interested' or one who is always running away. On the other hand, if you have a guest from Holland, it would be just the other way. You would be doing the chasing while your guest would be struggling to maintain a distance. This struggle is because your social space overlaps his personal space.

Haptics

Haptics, the sense of touch, is crucial for human exploration and manipulation of the world. However, haptics customs differ significantly throughout the world. For example, in many Latin American countries like Mexico, people greet each other with one kiss on the cheek, while in Spain, people greet each other with two kisses on the cheek.

British are usually not very physical. On meeting someone for the first time, the British would generally shake hands in a formal situation or even smile at each other. If it is a friend or casual acquaintance, they will hug or (between two women or a man and woman) make one kiss on the cheek. At the same time, Brazilians are well-known for the warm, Latin-American-like greetings, very demonstrative, festive with lots of kisses and hugs. Among men, if they are friends, there is generally a light hug and a tap on each other's back. Among men and women or women/women, kisses are the norm (Arena, 2010).

In the United States, the main form of greeting is shaking hands, though it is not uncommon for them to hug when greeting friends. In Japan, people bow to each other in greeting and think it rude to touch. In Thailand, also they do not touch, hug or kiss each other. However, instead, they greet friends and colleagues and pay respect to seniors by *'wai,'* which is a gesture where put their palms together in a lotus bud at their chest and bow their head so that the tip of thumb fingers touches their chin (friend), nose (senior), and in-between eyebrows (monk, royal family), almost like *'namaste'* in India. In Bangladesh and Pakistan, one makes a relaxed salute with the right hand (*Salaam*).

Every country/culture has a different perception of touch as communication. Therefore, as a tour guide, one needs to be careful.

Kinesics

Body language, also called kinesics, is the discipline concerned with studying those bodily motions that are communicative (Morain, 1978). Kinesics includes posture, movement, facial expression, eye management, gestures and distance. Cultural norms regarding kinesics vary considerably between countries; if you cannot read the guest's body language properly, you might be misunderstood and sometimes misinterpret the local non-verbal signals. This can undoubtedly lead to embarrassing and uncomfortable situations.

There are clues that guides may learn over time. Try to research where the customer groups are from and what they do. Some interesting examples include:

- The `O.K.' and `thumbs up' gestures are considered vulgar in Argentina
- Slapping your open palm over a closed fist is considered a vulgar gesture in France.
- Pointing with the index finger is considered rude in Japan. You should point with the entire hand as also in China.
- Staring is considered rude in Britain and makes people feel uncomfortable.
- Do not stand with your hands on your hips; this signifies anger. It is considered rude to stand with your hands in your pockets in Mexico.
- Do not touch the head of a Thai or Cambodian.
- The Chinese do not like being touched by strangers. Do not make any body contact.

- Direct eye contact and staring are uncommon in China when walking in public places.
- Americans are often uncomfortable with same-sex touching, especially between men.
- Nigerians do not use the left hand to give or receive objects, as the left hand is considered unclean.
- When a Russian touches another person during a conversation, it is usually a sign of confidence. Speaking or laughing loudly in public is discouraged.
- Pointing the bottom of your feet at someone is an insult in Thailand and for Arabs.
- Turks will stand close to you while conversing. Do not back away, as this can be construed as unfriendly.

Vocalics

Vocalics is a part of non-verbal communication because it is not related to the content or verbal message but the other attributes of speaking, including the pitch, tone, volume, tempo, rhythm, articulation, resonance, nasality and even the accent of the speaker collectively known as *prosody*.(Andersen, 1999) For example, whistling is used for different purposes by people in different cultures – some use it to entertain by whistling a song. In contrast, in some cultures it is used to chase women on the streets because they appear attractive.

A *tisk tisk* in America expresses disappointment, but in some cultures can be an expression of attraction to someone else. Other sounds can mean the same thing in other cultures. For example, a *gasp* expresses shock or surprise, while a sigh typically expresses disappointment or fatigue (Fevgroup, 2007).

Religion and differences

Religion is one element that affects culture and is tangibly distinguishable in most cases. It is a codified belief system that allows or disallows certain things. There are several tangible elements like places of worship, dress code, symbols and motifs, and architectural elements. Then there are intangible elements like rituals and practices. All of these elements contribute to cultural distinctiveness.

Both culture and religion are human constructions. Each has its symbols and manifestations. Over time the distinction between religious and cultural seems challenging to maintain. It makes more sense to think of religion as a sub-system of some cultural systems, but culture can exist outside of religion. A religion can colour the local culture, but religion can also transcend different cultures. Culture is a way of life; religion is a belief system.

Religion tends to be codified. It involves rituals, books, or activities written down or taught in a specific form and preserved more or less the same. Religion, once codified, changes slowly. Culture, on the other hand, is generally not codified, but it can be described, and it changes more quickly in response to external environmental forces. One such force is often the influence of religion.

Often the line between religion and culture is fine. However, it is not easy to distinguish between a religious and cultural element in practice. The two are often combined. Further, these could be sensitive issues for an individual. Tour managers are faced with many such situations. There may be individuals in the group who are religious and submit themselves to certain kinds of religious-cultural restrictions – many Indians are vegetarians; Muslims do not consume pork, while Jews will have only kosher meat. A tour manager must be aware of his groups' religious preferences. They must be honoured, and personal space for adhering to religious customs may be provided to an individual as far as possible. Group, on the other hand, as part of their itinerary, would be visiting many religious places – churches, temples, mosques, or synagogues. The tour guide must be aware of the sensibilities of these religious places and must advise the group accordingly. For example, he must advise them on the proper dress while visiting a church or a mosque.

Valuable tips to managing the cultural differences

- Observe facial expressions.
- Watch the body movements of your guests and try to decipher what they mean.
- Watch for and interpret vocalics like laughing, coughing, yawning, etc.
- Try to learn about the differences between your guest's social, political and economic systems and yours.
- Try to make sense of how these differences affect his/ her thinking, ways of learning.
- Know your tourist's legal, technical and financial systems
- Know how the above will affect your choice of tactics
- Know the effects of ethical standards (right, wrong, reasonable)

Key points

Tour guiding is essentially about communication with the visitors to help them negotiate the place with understanding. Therefore, good communication is the foundation of effective interpretation. A tour guide will often interpret gaze in different situations – a historical site, a walking tour, a church, a museum, or a natural setting. However, the primary communication principles will remain the same, and a tour guide must hone interpersonal communication skills.

Communication has two ends – the sender and the receiver. In a touristic interpretation scenario, a tour guide is the sender while the visitor is the receiver. The visitors must receive in totality what is communicated by a tour guide. Often some distractions and barriers distort the communication. The effectiveness of communication also depends on the needs of the receiver, the visitor.

The most accepted communication process model includes sender, receiver, encoding, decoding, message and feedback. Then there are various channels or mediums of communication like audio, video, electronic channels, written or body language.

In our case, a tour guide is the sender or initiator of the message. Guests trust the message sent by tour guides only when they consider them trustworthy or credible. Credibility is created through three elements: the message's content, the tour guide's confidence, and commitment. Eye contact with the guest, body language, tone of voice and smart appearance of the tour guide are ways to appear confident.

The receiver of the message, in our case, is the visitor. If the receivers are not appropriately situated, they may not receive the message correctly. In addition, there may be external distractions like noise or some commotion. Further, a tour guide must also understand the communication style of the guest so that they may adapt their communication accordingly. Four personality types have been identified – thinkers, feelers, intuitors, and sensors.

The tour guides need to understand the differences in the background from which an individual comes. At least four primary ways a visitor confronts an alien culture are time, space, context, and religious beliefs. Time is understood in terms of monochromic vs polychromic cultures. Space can be understood in terms of proxemics (use of space and proximity), haptics (touch), kinesics (body movement) and vocalic (paralanguage). In terms of context, one could be from high context and low context cultures. Finally, it is also essential for a tour guide to understand the religious differences. Valuable tips are given so that the tour guides can excel in their communication.

Key terms

Communication The exchange of information by speaking, writing, or using some other medium from one place, person or group to another.

Decoding The process of receiving, interpreting and understanding an encoded message.

Encoding The conscious act of turning thoughts into communication by using a medium like a phone call, email, text message, face-to-face meeting, or another agency.

Feelers Personality types who have a more personal communication style. They process information and draw conclusions based on their emotions. They 'feel their way'.

Haptics Haptics is a form of nonverbal communication about how people and animals communicate and interact via the sense of touch.

High context Refers to cultures that are usually relational and collectivist. It assumes the individual is knowledgeable about the subject and has to be given very little background information.

Intuitors	Personality types who do not deduce a conclusion logically; instead, they draw intuitive conclusions. Intuitors do not know how they arrived at a conclusion yet believe they made a correct decision.
Kinesics	The study of nonverbal communication concerned with interpreting body movements, postures, facial expressions, and eye contact.
Low context	Refers to cultures assume that the individuals know very little about what they are being told and must be given a lot of background information. Therefore, communication depends more on spoken words than interpreting more subtle or unspoken cues.
Message	A piece of information conveyed by words (in speech or writing) and/or other signs and symbols; the content of the communication process. It could be verbal or nonverbal, or both.
Monochronic	Refers to a cultural type where time is considered linear and a valuable commodity that should not be wasted; therefore one should engage in one thing to ensure that time is well-managed.
Polychronic	Refers to a cultural type based on time orientation where time is considered spiral, flexible, and abundant. Polychronic individuals are multitasking and can execute several tasks with a natural rhythm.
Proxemics	The study of how humans use space in communication. It focuses not only on how people orient themselves to other people and objects in their immediate physical environment but also on these orientations' perceptual and behavioural impact.
Receiver	The individual or individuals to whom the message is directed.
Sender	The originator and despatcher of a message sent to a receiver through a particular channel on a particular occasion.
Sensors	Personality types who use the input from their senses to inform them. Sensors are forceful, direct, poor listeners who look for quick direct talk with few details.
Thinkers	Personality types who analyse consciously and deduce information. They are comfortable with the use of data that includes facts and figures. They continuously seek the logic behind everything being said.
Vocalics	The study of paralanguage, which includes the vocal qualities that accompany verbal messages, such as pitch, volume, rate, and verbal fillers.

■ Discussion questions

1. Would you agree that communication is one of the essential skills for a tour guide? Elaborate your reasons.
2. How does the understanding of Maslow's Need Hierarchy help you communicate with visitors as you lead them around a monument?
3. You are leading a group of visitors from the UK at a busy market, alive with a lot of the hustle and bustle, hawkers selling snacks and a lot of pedestrians and commuters. As a result, the noise level is unusually high. What are the likely challenges as you guide and instruct your group?

4 The guests' culture, gender, nationality, or social class will affect how they receive and interpret a tour guide's commentary. Discuss.

5 Discuss the different personality types of guests. What would be your strategy to communicate with them?

6 A tourist at immigration gets frustrated as the customs official takes too long to stamp her passport. He then takes some more time to indicate her to pass through the body scanner. The official is slow as he keeps talking with other colleagues, and he takes calls as his phone rings frequently. On the other hand, the tourist wants the customs official to attend only to the job at hand – stamping her passport – before moving on to the next. Instead, the official attends to many tasks simultaneously. How would you explain this?

■ Activity

1 Sit in groups and assess mistakes that a tour guide can commit while communicating and highlight the key points.

2 Find out the best exercise for communication that a guide can undertake before starting a tourist guiding assignment.

References

Arena, C. (2010). Kiss, hug or shake hands? Retrieved May 18, 2020, from Pocket cultures. https://pocketcultures.com/2010/07/14/kiss-hug-or-shake-hands/

Andersen, P. A. (1999). *Nonverbal Communication: Forms and Functions.* Mountain View, CA: Mayfield.

Brownell, J. (1982). Increasing your credibility. *Journal of Supervisory Management, 27*(12), 31-36

Dictionary.Com. (2021). *Eye Contact.* Retrieved 1 8, 2021, from Dictionary.com: https://www.dictionary.com/browse/eye-contact?s=ts

Fevgroup. (2007). Culture and nonverbal communication. Retrieved Nov 2012, from https://fevgroup.wordpress.com/2007/09/18/culture-and-nonverbal-communication/

Guerrero, L.K., Devito J.A. & Hecht M.L. (1999). *The Nonverbal Communication Reader: Classic and contemporary readings* (2nd ed.). Long Grove, IL: Waveland Press, Inc.

Hamermesh,D.S. (2011). *Beauty Pays*, Princeton University Press.

Jung, C. (1913). *Psychology Of The Unconscious.* London: Kegan Paul Trench Trubner.

Jurin, R., Danter, K., & Roush, D. (2000). *Environmental Communication: Skills and Principles for Natural Resource Manager, Scientists and Engineers.* Boston, MA: Pearson Publishing.

Maslow, A. H. (1943). A theory of human motivation. *Psychological Review, 50*(4), 370-96.

Mehrabian, A. (1981). *Silent Messages: Implicit Communication of Emotions and Attitudes* (2nd ed.). Belmont, CA: Wadsworth.

Merriam – Webster. (n.d.). *Communication.* Retrieved January 6, 2021, from Merriam Webster: https://www.merriam-webster.com/dictionary/communication

Morain, G. G. (1978). *Language in Education: Kinesics and Cross-Cultural Understanding.* Center for Applied Linguistics.

Pachter, B. (2013). *The Essentials of Business Etiquette: How to Greet, Eat, and Tweet Your Way to Success.* New York: McGraw-Hill Education.

Ruth, U. (2017, 8 17). How making eye contact can help you appear more confident at work. Retrieved 1 7, 2021, from Make It. https://www.cnbc.com/2017/08/17/how-making-eye-contact-can-help-you-appear-more-confident-at-work.html

TIC. (2021, 1 7). *Tourist Guide.* Retrieved 1 8, 2021, from Travel Industry Council: http://www.tichk.org/public/website/en/guides/code/html

Tilden, F (1967). *Interpreting Our Heritage,* University of North Carolina Press.

TravelWild. (2017, 1 3). *TravelWild Australia.* Retrieved 1 8, 2021, from Tour Guide Code of Conduct: https://travelwild.com.au/employment/tour-guide-code-conduct/

4 The Art of Interpretation

Learning objectives

After studying this chapter, you should be able to understand:

- The nature of interpretation for tour guides
- The characteristics of a powerful interpretation
- The TORE model of interpretation
- The techniques of interpretation

As we often say, tourism is no longer about seeing a place – it is about experiencing it. Tourists like to have a feel of the place and get immersed. Naturally, they would like to understand, especially when travelling to a very different place. However, they often fail to make sense of animate and inanimate elements of this alien space. For a long time, tour guides have been providing hordes of monotonous information about monuments and places, making it difficult for the tourists to appreciate and therefore connect to the object, the event or the activity.

Tour guides are expected to put meaning to the place they show around. They are, therefore, also *interpreters*. This approach has dictated a change in their roles as well. Information and interpretation are not the same things. Information is simple straight facts and figures presented to the visitors, whereas interpretation adds meaning to the information. According to John Veverka and Associates, "visitors travel to see, experience and learn (edutainment) about natural or cultural landscapes, sites, features, objects, people, events and stories. It needs to be noted here that the educational component (learning and connecting) of this type of tourism is the key aspect. Visitors want to learn, see, and do! They travel to sites for a mix of edutainment experiences." (Veverka, n.d.)

Interpretation

Interpretation is not about translating information from one language to another. According to Dr Sam H. Ham, each audience will eventually develop their understanding of the subject and, therefore, will interpret it differently. Therefore, professional interpretation by guides is a process of provoking audiences to think. The job of a guide is to engage visitors through their storytelling so that they are hooked and make conscious efforts to construe the meaning.

Interpretation communicates what is noteworthy about places, people or events – why they are appreciated? It is a celebration of place and things, culture and nature, creativity and folklore, great successes and failures in history. Interpretation is connected to sites and objects, artwork or living things, and it can happen anywhere; in parks, visitor centres, historic sites, city streets, museums, zoos or galleries, at special events or promotions, and in publications (Colquhoun, 2005). Interpretation, therefore, is a communication process designed to reveal meanings and relationships of our cultural and natural heritage to visitors through first-hand experiences with objects, artefacts, landscapes, and sites.

Example of interpretation:

Fact: Plants and animals are worshipped in India

Interpretation: This is the Indian way of appreciating nature; due importance has been associated with plants, animals, and other ecosystem constituents to be venerated, preserved, and protected.

Characteristics of a powerful interpretation

The eyes are a window that leads to the mind. They focus on what is at the centre of a person's visual attention. However, the visual stimulus becomes information when associated with a cognitive process. Starting with a view, the visitor tries to gaze. To gaze, he/ she tries to add meaning to what is visually available. This is called the 'tourist gaze' (Urry, 1990). A tourist's desires to visit places and the ways we learn to appreciate them visually are not merely individual and independent – they are culturally mediated and socially organised. Tourists depend on visual and narrative depictions to make sense of the place. This sense-making is called interpretation. In simple terms, the tourist experience of a place involves a particular way of seeing. He wants to add meaning to what he sees. The visitor needs assistance to interpret the place and its elements. Assisted mediation builds narratives about the uniqueness of their history, culture and heritage, and varieties of performative practices help the tourists articulate experiences.

A powerful interpretation of place, things and events is the soul of good tour guiding. A good interpretation

- Excites the attention of the visitors – it emphasises why a (cultural or natural) site has value to them (guests), community and to the world.
- Relates the attraction of a place to the everyday life of the host community (their attitude, viewpoints and philosophy of life).
- Motivates the visitor. They come for the story with the site as the contextual backdrop. Without interpretation, it is just another old structure (cultural site) or a part of the place (natural site).
- Establishes the importance of the site, attracting attention and concern for it. Conservation, preservation and restoration follow logically.
- Customises the experience and the story for the visitors. Each guest is an unique person. Thus, a good interpretation may vary to attend to different tourist groups' interests, backgrounds, feelings, and expectations.
- Makes the performance of a tour guide enjoyable, inspirational, impressive and persuasive.

Tilden's six principles of interpretation

Freeman Tilden (1883-1980) worked with the United States National Park Service. He was the pioneer of interpretation for visitors. In 1957 he authored a book *Interpreting Our Heritage*. His oft-cited quote is, "Through interpretation, understanding; through understanding, appreciation; through appreciation, protection." He gave six principles of interpretation (Tilden, 1957).

1. Any interpretation that does not somehow relate what is being displayed or described to something within the personality or experience of the visitor will be sterile.
2. Information, as such, is not interpretation. Interpretation is revelation based on the information, and they are entirely different things. However, all interpretation includes information.
3. Interpretation is an art, which combines many arts, whether the materials presented are scientific, historical or architectural. Any art is, to some degree, teachable.
4. The chief aim of interpretation is not instruction but provocation.
5. Interpretation should aim to present a whole rather than apart and must address itself to the whole man rather than any phase.
6. Interpretation addressed to children (say up to the age of 12) should not be a dilution of the presentation to adults but should follow a fundamentally different approach. To be at its best, it will require a separate program.

Types of interpretation

There are two types of interpretation – personal and nonpersonal. Personal interpretation is where tour guides or interpreters interpret a place or object during a physical face-to-face interaction. Nonpersonal interpretation is narrative depiction through exhibits, audio-visual aids, printed material, and other means without mediation by a guide. Thus, the consumer is in control of nonpersonal interpretation, for example, they can go forth and back on audio-video aids, selectively read the exhibits and brochures and vary the sequence as they choose. On the other hand, the interpreter leads the discourse in personal interpretation, and is mainly responsible for his commentary, contents, screenplay, and adaptation to the situation at hand. The visitor can be a simple spectator, or the interpreter may encourage more active engagement of the audience.

A comprehensive interpretation plan comprises personal and nonpersonal interpretations used in tandem. Therefore, providing nonpersonal interpretive tools to supplement the personal interpretation by a guide at a site is an integral part of any comprehensive interpretation plan.

Nonpersonal interpretation

Nonpersonal interpretation is often used to interpret nature and culture. These self-learning materials help the visitor make sense of the pace and its elements. It is often used to direct, orient, and educate the visitors without a role for staffed interpreters. There are several tools used for nonpersonal interpretations:

People. Staff are sometimes used to play roles to demonstrate a phenomenon or concept. The actors in character are costumed to depict a historical or contemporary period. For example, at Shakespeare's birthplace at Stratford-upon-Avon, costumed actors perform snippets from his famous plays.

Figure 4.1: Glimpse of a Shakespeare play at the museum in Stratford-upon-Avon. Picture credit: Prof. Alok Chakrawal

The Nederlands Openluchtmuseum is a cultural-historical museum and gives a perspective of the Netherlands through the centuries. When visitors step inside the museum, they are taken on a journey through time, from two centuries ago to until recent times. The open air museum brings history closer through authentic buildings, objects, and stories. The costumed staff showcase the life and times of early Netherlanders. The staff seem to have walked out of the past. Proud of their heritage, they inspire visitors with their stories.

Media. These nonpersonal interpretive tools includes virtual environments, interactive devices, educational signage, graphics, sound, smell, and interactive or static exhibits. For example, many natural and cultural attractions introduce visitors to the place through videos played in visitor information centres. These materials are often available as souvenirs. Sometimes, radio transmissions are used to reach visitors' vehicular radios in outdoor environments. Roadside displays help visitors to tune in.

Objects. Sometimes, three-dimensional models, objects, or artefact-based displays are used for nonpersonal interpretation. Buildings, stuffed or real animals, plants, rocks, models, figures, and props are often used as exhibits to add meaning. Exhibits can be interactive and can include written, visual, kinesthetic, or auditory methods of communication.

Experience is incomplete unless one uses all the senses, including touch. It is a natural impulse to touch things to feel them. Traditionally, visitors have been advised not to touch artefacts and exhibits. However, this is changing. For visitors to make sense of what is being displayed, exhibitors and curators are now using technology to allow people to feel the exhibits. Manchester Museum uses these technologies, enabling its visitors to interpret the exhibits.

One such technology is digital touch replicas, also called digitally scanned replicas. It involves accurate scanning and replication technology which will reproduce near-original shape and surface. Visitors, especially those with sensory impairments, can engage with replicas. Signage and tour guides encourage visitors to feel the artefact for themselves, allowing a more profound sensation. The University of Manchester used this technology to produce a Confuciusornis Sanctus fossil feather replica. Manchester museum used another technology - the Probos Sensory Consoles, to enable visitors to touch and interact with digitally scanned museum artefacts that would otherwise not be touched because they are too delicate or protected in a case.

In 2014, the Metropolitan Museum of Art in New York created 'Multisensory Met'. This innovation could add sound and smell to miniature replicas of the museum's famous artefacts to bring them to life. Similarly, the Tate Britain in London used 'Sensorium'. It recreated tastes, sounds, and smells, and could make specific colours appear intense

Self-guiding tools. There is increased use of technology to empower visitors on self-guided tours. Conventionally, trails were interpreted using brochures, signs, and wayside exhibits located alongside the trail, road, rail, and even underwater and in the sky. Of late, podcasts, cell phone tours and media have been deployed to augment visitors' understanding of the place on their own. Interactive kiosks may be placed for visitors to retrieve information as required. Attractions are also using GIS-enabled tools for self-guiding.

Built edifices. Comprehensive structures can be created to interpret a concept. For example, the Indira Gandhi Rashtriya Manav Sangrahalaya in Bhopal has been developed to present an integrated story of the evolution of man and culture with particular reference to India. At a nominal level, interpretation centres are developed to help visitors understand the place.

Figure 4.2: Indira Gandhi Rashtriya Manav Sangrahalaya, Bhopal. Picture credit: Vineet Saxena

Published material. Nonpersonal interpretation material like brochures, pamphlets, and maps, among others, are the most common tools for interpretation. These publications are frequently used to inform and educate visitors.

Signage. Signage is yet another essential tool used for interpretation. Freestanding signages are the most commonly used vehicle to communicate with visitors, with some bearing details about the displayed artefact, structure, or place in almost every case. Nowadays, signages are produced using a range of materials. Some signages could be interactive also.

There are a few limitations of nonpersonal interpretation. One, the visitors have the freedom to choose. They may also have to decide the sequence in which they would interact with the artefacts and exhibits. Since they are new to the place, they may miss the underlying story. They may not know what is essential and what is not. Though the curators may suggest a sequence or a route, in the absence of guidance, the visitor may find it difficult to relate. Visitors often look for expert advice while visiting an unknown place. Two, the inappropriate use of nonpersonal interpretive tools may cause distraction. Often the exhibits and signboards are used far too long, and they get worn out. It could be an eyesore in an otherwise visually appealing view of a place. Three, the use of technology can be costly, and so also its maintenance. Recovery of costs of nonpersonal tools could be a consideration for the decision-makers at a destination. However, in some economies, the cost of engaging a tour guide may be high, which could encourage investment in nonpersonal interpretation. Four, not everyone likes reading. They might not like to invest much energy in interpreting the tourism product, and their interest might fade.

Sometimes, tour guides and interpreters consider nonpersonal tools as adversaries that may substitute them. However, this is not the case. Humans have an innate craving to connect with others for mental and physical wellbeing. There is compelling evidence to support this theory. A part of the motivation for visitors is to see new people and broaden their horizons. It is about the interaction with the people they meet. Individuals become desensitised, numb, and oblivious to the social cues in the absence of human interaction. They, therefore, seek to accompany locals to make sense of a new place. It creates a social vibe. Therefore, tour guides and leaders should consider the nonpersonal interpretation tools as supports and include them in their overall interpretation plan.

Personal interpretation

Personal interpretation is face-to-face interaction with visitors. It is available in various formats – guided tours, walks, talks, demonstrations, and others. One of the critical characteristics of personal interpretation is that the tour guides are mostly in control. They decide the content and sequence of flow of information and even the routing. What follows what and what is to be said at each point is decided by the deliverer. They can even modify the narration to suit the situation. Further, care and attention are an integral part of personal interpretation.

Guides, the experts who control the message, can highlight what is noteworthy. They also facilitate the visitor to immerse in the authentic experience of the locale. Equipped with multiple options for conveying the message, they use the best as the situation demands. Spontaneous responses to their queries further enhance the experience of the visitor. They answer questions, clear misunderstandings, add meaning to what is present, and advise how to make most of the situation. They shape personal experiences, which can be rewarding and satisfying for the guests. Further, often being the local, the tour guide adds an 'authentic' to the experience. They also lead the visitors to physically engage in the locale to augment the authenticity of the experience.

How to interpret?

Dr Ham suggested a four-pronged model that we will refer to as the TORE model (Ham, 2013). This model suggests that any presentation by the tour guide must comprise of four essentials:

1 There needs to be an underlying theme (T) in the narration.
2 Commentary of a tour guide needs to be organised (O)
3 It needs to be relevant (R)
4 The narration needs to be enjoyable (E) for the audience.

Interpretation should have an underlying Theme (T)

The information presented to visitors should have a theme. The theme makes the job easier for the tour guide. It guides them on what to include and what not to include in their narration. It also suggests what elements are to be emphasised. The theme is not the same as the topic or subject of the information. Topic refers to the subject matter, whereas a theme is about an underlying story that you would like to tell depending upon the taste and interest of the listener. It is about a specific idea or the main points you would like to share. Thus, a tour guide must consciously decide what is to be shared with the group out of all the information that they have about a monument or event or any other attraction.

One of the mantras is to limit the key concepts that you present so that the guests can assimilate them. If the number of concepts is large and varied, guests are likely to miss a couple of them and gradually, the whole interpretation hazes away. It is advised that the essential concepts associated with the interpretation of a site, event or object should not be more than three to four. It will increase the likelihood of the interpretation being understood more successfully. The second mantra is to weave these three or four concepts into an exciting story. Stories are easily remembered, and the interpretative argument will appear more reasonable and therefore organised.

While tour guides are taking a group around the Pink City Jaipur in Rajasthan, India, at Jantar Mantar, they may choose a theme to speak. While the topic would be Jantar Mantar, they may exercise their ingenuity to speak on one of the many themes. For example, they may speak about the precision of astronomical calculations using the Jantar Mantar observatory, or the architectural design of the structure and various *yantras* (equipment) there; alternatively, they may speak about the history and need for astronomical directions in the contemporary period. So, in this case:

The topic could be:	Jantar Mantar
The theme could be:	The need for astronomical directions in the contemporary period
The talking points:	1. Raja Jai Singh of Jaipur and his love for astronomy
	2. Five sites of observatories in India
	3. Major instruments created
	4. A glimpse of what could be achieved with these installations in those times

The theme should connect the underlying key, with a maximum of five key points. According to Sam Ham, interpretation of a site, event, or activity should try to answer the question 'So what?' When the key points converge to the same thematic idea, it helps the guest focus on answering 'so what?'

The theme of interpretation should:

- Be short, concise, specific and straightforward
- Have a focus (all underlying key points should converge to the theme)
- Try to answer the question 'so what?'
- Be attractively worded.

As another example, while the theme could be `*Maharaja Bhupinder Singh Ji was the most famous ruler of Patiala, best known for his lifestyle and extravagance*', four key talking points could be:

1 He was a great benefactor of sports. Owned cricket and polo teams 'Patiala XI' and 'Patiala Tigers' and captain of the Indian Cricket team.

2 He married at least ten times and had many consorts. Maharaja used his tactfulness to keep them all happy and content.

3 He was the proud owner of the world-renowned 'Patiala Necklace' made by Cartier SA.

4 He was the first man in India to own an aircraft.

Interpretation should be Organised (O)

If your is not information organised, the audience will be more often than not distracted by other things. Organised information is presented such that it is easily understood and easy to follow. The best interpretive moments are when the audience does not have to work hard to understand what is being presented. Also, it is easier to impart the information in a smooth flow with organised information. This approach is handy when the content is complicated. The art lies in presenting a complicated subject simply. There are a few things that the tour guide must keep in mind.

Avoid Too Much Information (TMI)

One could sometimes find a tour guide in Jaipur who would spend 15 minutes explaining one window frame on *Hawa Mahal* during a 2-hour city walk. Moreover, this deliberation could be with school kids who are in Jaipur for a day stopover. Do not exhaust guests with information overdose. Some people like to stuff listeners with their knowledge, opinions, and useless facts when they have a captive audience. It is truly a waste of time. This is TMI syndrome.

In her book *Le Jardin d'Épicure*, a famous literary figure of nineteenth-century, Anatole France has rightly advised, "Do not try to satisfy your vanity by teaching a great many things. Awaken people's curiosity. It is enough to open minds;

do not overload them. Just put there a spark. If there is some good inflammable stuff, it will catch fire". A reasonable interpretation provokes tourists' interest and attention to what may be an unfamiliar subject for them.

Principle of KISS

KISS is an acronym for "Keep it simple, stupid". The principle is attributed to the US Navy of the 1960s. The principle suggests that things should be presented as simple as possible. Also expanded as "Keep it short and simple", the principle applies to interpretation scenarios. While the tour guides should have extensive knowledge, the art lies in presenting the most useful and interesting facts in the shortest possible time and most straightforward possible language. Do not use clichés and technical jargon to make things difficult for the listeners.

Have an elevator pitch

An elevator pitch is a brief, persuasive monologue or prologue that a tour guide uses to generate interest in what is to be presented next. A good elevator pitch should not be longer than 30-50 seconds – the duration of a short elevator ride, and hence the name. The elevator pitch should establish what makes the artefact, its legacy, or its gaze unique. Always remember that timing calculations need to be in your head, and constant checking the watch while guiding is a big NO!

Unfortunately, some tour guides think their primary job is to provide information. No. Their job is to ignite curiosity. They must speak less. A tour guide must present the essential information in a way that excites their guests and provokes them to ask questions. Spend more time answering these questions.

Must tell – Should tell – Could tell

This popular tool is advised by experienced guides at WFTGA (World Federation of Tour guide Associations). It helps in organising the content of commentary in three components:

Must tell: this includes information essential about the artefact, monument, place or event. This part is like the elevator pitch. Talk about two things here: what makes the monument or event unique. What is the USP (Unique selling proposition)? Second, share some fundamental facts about the monument/ event. (Tip: spend no more than 60 seconds on this). This component generally includes who? When? Why? And sometimes, with what (material)?

Should tell: these are the next important things about the monument or the event. It provides context to what had been included in the 'must tell'and answers more, Why? What? Who? How? Where? It may include the famous tale about the monument/event. (Tip: devote no more than 90 seconds to this).

Could tell: may include possibly everything about the monument. Depending upon the time available and upon the willingness/eagerness of the guests, this section can be expanded or contracted.

Figure 4.3: Hawa Mahal at Jaipur, using the Must tell – Should tell – Could tell model.

What is the importance of this monument?	Who built it? /When was it built? / Why?	Architecture/How (with what material) was it built	Other interesting things
Must tell			
Called 'palace of wind or palace of breeze.' Signature building of the city	Sawai Pratap Singh Designed by Lal Chand Ustad It was built in 1799.	Five stories; Red and pink sandstone carefully and painstakingly outlined with white borders and motifs; 953 small windows called *jharokhas* decorated with intricate latticework	Jaipur's signature building
Should tell			
Pratap Singh's deep devotion to the Hindu god Lord Krishna is inferred to have prompted him to build it as dedication, in the form of a Mukuta or headgear, adorning the Lord.		It rises to 50 feet (15 m) from its high base. It forms part of the City Palace and extends to the Zenana or women's chambers	It is particularly striking when viewed early in the morning, lit with the golden light of sunrise
Could tell			
It is built in red and pink coloured sandstone, in keeping with the décor of the other monuments in the city; its colour is a complete testimony to the epithet of 'Pink City' given to Jaipur	Maharaja Sawai Jai Singh, the ruler of Rajasthan of the Kachwaha clan, founded Jaipur city in 1727. His grandson Sawai Pratap Singh, son of Maharaja Sawai Madho Singh I, built the Hawa Mahal in 1799 The lattice windows allowed royal ladies to observe everyday life in the street below without being seen since they had to observe strict purdah.	The top three floors of the structure have a dimension of one room width, while the first and second floors have patios in front of them, on the rear side of the structure. As seen from the street, the front elevation is like a honeycomb web of a beehive built with small portholes. Each porthole has miniature windows and has carved sandstone grills, finials and domes It is a veritable mass of semi-octagonal bays, which gives the monument its unique façade.	The royal family of Jaipur, during their reign, used the palace as a hot weather retreat, during the suffocating summer season of Jaipur, since the unusually designed window screens provided the much needed cool breeze and ventilation. A fusion of Hindu Rajput and Islamic Mughal architecture; the Rajput style is seen in the domed canopies, fluted pillars, lotus and floral patterns, and the Islamic style is in its stone inlay filigree work and arches

■ Interpretation should be Relevant (R)

Relevant means two things. One, it must be *meaningful* for the audience and two, it must be *personal*. Meaningful means that the interpretative narration should bring in new ideas, link them, and build them upon what is already in the minds of the audiences. This 'connect' is best achieved through examples, analogies, contrasts, similes or metaphors. For example, a tour guide may ask the audience to imagine how difficult it would be if we did not know the time and how would people know about the time of the day when there were no watches. This activity may help guests stretch their imagination and relate to the importance of this mammoth structure built almost three centuries back.

Another aspect of being relevant is that the interpretation should connect them to something they care aboutl. Interpretation tries to relate the site or artefacts to visitors' knowledge, experience, background and values. That is where it becomes personal. While our standard watches measure up to one second, instruments at Jantar Mantar are intended to measure up to half a second. It is an equinoctial sundial. On one side of the gnomon is a quadrant of a circle parallel to the equator's plane.

The key is to connect with the audience as quickly as possible. See what is important for the visitors and connect to that.

Making interpretation meaningful

Excessive use of technical jargon will quickly kill interest. Instead, consider the following:

Use illustrations

Try to show your visitors an example that represents the concept you want to convey, such as, "This is betel quid called '*paan*' which are prepared with slaked lime, fennel seeds coconut, fruit preserves, rose petal preserves called *gulkand* and various spices. Variants of this are used in different parts of Asia. This chew is often used as a mouth freshener".

Sensate

If possible, demonstrate an effect. Make the guest feel for themselves. Also, add an explanation. For example, "Come here and stand in front of this stone grill window! Feel the cool breeze. As air passes through these cuts, they serve as venturi and expand the air suddenly and cool it."

Juxtapose

Draw comparisons from what is already known. These comparisons will help guests conceptualise and remember the facts and help them create mental pictures by drawing comparisons with everyday objects as references. This comparison will help tourists understand and retain concepts with much ease. For example:

Ladies and gentlemen! Here we are in the *Nanakshahi* brick fort of Malerkotla built by Nawab Bayazid Khan during the 17th century. *Nanakshahi* bricks were named after Guru Nanak, the founder of Sikhism. The *Nanakshahi* bricks were less than ¾ in every dimension from the bricks used in India nowadays. They are thin bricks of approx. 6″ × 4″ × 1.25″.

Tour guides may also highlight the difference "this fort is built on the plain unlike all other forts built on the top of the hill."

Time hooks

The guests will find it difficult to remember the dates that tour guides tell them. So many dates and so many names are confusing. Guests might not be interested in exact dates. However, they would like to have a general idea of the antiquity of a place or a historical event. A smart way is to hook the antiquity of the place or event with some event or monument from the visitor's context. Say, for example, if a tour guide is handling a group of Chinese guests, while speaking about Taj Mahal, they may point out:

> "Mughal ruler Shah Jahan built this magnificent monument almost the same time Ming rule was coming to an end".

If the guests are from Britain, they may tell the group:

> "Taj Mahal was completed in the year Oliver Cromwell died".

Or for example:

"When Armenia became the first Christian state, Vatsayan in India was writing *Kama Sutra.*"

Build links

Links are similar to hooks in that they try to link the object being observed to what they have seen earlier on the trip. Links could be contrast or parallels. For example, tour guides need to know where tourists have been and what they have seen. You can then link your story to the previous one. For example:

> "Ladies and gentlemen, we stand before the marvellous Taj Mahal, which was built by the great Mughal Emperor Shah Jahan, who also built the Red Fort and Jama Masjid in Delhi, which you visited the other day."

Draw parallels

Comparisons make it easier to comprehend the underlying concepts and sometimes the philosophy. Draw analogies to something the guests are familiar with.

> "Chittorgarh Fort is the most massive fort in India and the grandest in the state of Rajasthan. The fort precinct with an evocative history is studded with a series of historical palaces, gates, temples and two prominent towers. Like most medieval castles in Britain, it is surrounded by a tall curtain wall, parapet wall with brattice and corner watchtowers."

or

"These Hindu priests can marry and lead a family life as many Eastern parish clergies."

Do not trivialise

Some tour guides mistakenly believe that their job is to deliver long narrations. They often run out of content. They stretch it with trivia making it difficult for the guests to select and pick up helpful information. As suggested above, a tour guide has minimal time, say 30 seconds to 2 minutes, to say it all. One must be very picky about information and words. Precise words and concepts, short sentences, and simple language is the key. Do not waste time with trivial information or ubiquitous things.

"People walk on this street" (People everywhere walk on the street)

Making interpretation personal

Making the interpretation personal means helping visitors draw parallels from their own lives and lifestyles. This tactic will help a tour guide immediately capture the guests' imagination who connect it to self as they provide new information. This technique is also called "self-referencing".

"How many of you can build a bridge for your girlfriend? Once, a young Muslim dancer (*Kanjri*) Moran, hailing from Lahore, was on her way to dance at the Maharaja's Ranjit Singh's Baradari. Her shoes fell into the water channel on the way. The piqued dancer refused to dance until a *pul* (bridge) was built! This place has become popular as Pul Kanjri".

"What you just spent on that Coke is what this family has to survive a day."

Interpretation should be Enjoyable (E)

The bottom line is that visitors have come to have fun. It cannot be an uninteresting classroom discourse. The last important element of thematic interpretation is that it must be enjoyable, pleasing and engaging.

Principle of recall

The formula for the recall is simple. People will remember-

- 10% of what they read
- 20% of what they hear
- 30% of what they see,
- 50% of what they see and hear,
- 70% of what they say, and
- 90% of what they say and do

The message is clear. As tour guides, do not depend much on what the guests hear. Commentary should complement what they see. Good commentaries will force them to think and ask questions. They will remember answers to their questions more than what they had been told unsolicited. Allow them to 'do'. Allow your guest to explore the monument on their own – appreciate it and raise queries, have answers and be satisfied.

- Involve the group in *activities* (active learning). Thus, it will be a good idea to engage the group at Jantar Mantar to experiment with the basic time calculations that can be engaging and entertaining for them.
- Use *active verbs*. 'Active' language (instead of passive) is a powerful way of bringing presentations to life. In India, when tour guides translate from Hindi, they invariably use a passive voice. For example, instead of saying, "These magnificent temples were built by the kings to show their might", a guide may say, "kings built these magnificent temples to show their might". (We discuss this in detail later in the chapter).

Painting mental pictures

Often, tour guides and guests are at a place that was used differently in the past or is used differently on some occasions. Therefore, one of the more powerful tools of the tour guide's narration is his ability to paint a picture, making the audience feel as if he/she is witnessing an event (of past).

> "As the victorious king, would return to the palace riding on the back of an elephant, the rejoicing public would watch....and as the king's elephant would pass through this magnificent gate made of wood and iron in front of you while his queens would watch him from the oriels (*Jharokha*) above and shower petals .... the courtesan would occupy her place in the two baldachins (*Chattari*) on either side and dance while the musicians would play the tabor (*Tabala*) and zither (*Sitar*)...."

Interesting folklore/stories

Tourists are there to have fun. Lengthy commentaries with too many details can be dull. The audience may not be able to retain all the information, and therefore it is always a good idea to include folklores and exciting stories about the monument/period. This folklore helps tourists to create mental pictures and connect with the place. Exciting stories are easier to remember. However, the tour guide must always declare that these are stories. They should never present fiction as facts. Accomplished tour guides spend much time researching and collecting stories and tales about the place from various sources – libraries, archives, community elders, and others. You may also mention the shooting of movies and the stories associated with it.

Histrionics

Use of some histrionics is advised. Histrionics refer to a dramatic performance. Though exaggerated dramatic performances may be avoided, subtle facetiousness – voice modulation, pitch and pace variation may make commentary exciting and powerful. Disproportionate histrionics, however, will turn a tour guide into a distraction, which should be avoided.

The Dancing Guide of South India

Nagendra Prabhu, a tour guide in the temple city Madurai, recently rose to fame for his knowledge and dancing skills to interpret *Bharatnatyam*, a classical temple dance from South India. Prabhu describes the temple's architecture, paintings, and history to foreigners and domestic visitors who arrive to see the city. He uses his expertise to create various *Mudras* to keep his guests engaged and entertained. Mudras include poses of dance, hand gestures and facial expressions. Prabhu explains the meaning of the mudras or postures as he performs. He often depicts waves, a cobra, lotus, and peacock postures. He also tries to lift his leg to show the *tribhanga* posture of Lord Shiva.

He even invites his guests to follow his facial expressions and dance poses. It is an entertaining photo opportunity for the guests who participate enthusiastically. Several videos of guests performing with Prabhu are circulating on social media.

If he gets to know that the group is next travelling to Kerala, another south Indian state, he performs to explain to them the difference between Bharatanatyam of his state Tamil Nadu and Kathakali of Kerala.

Photo credit: Nagendra Prabhu

Nonpareil content

Over time, the narration about places and monuments gets clichéd and boxed. It is available in short guidebooks or hearsay that passes from one tour guide to another. If everyone says the same thing, a visitor may choose anyone. As an entrepreneur, a tour guide must try to differentiate himself/herself from others. Each guide should add well-researched information or a different perspective to the content of their narrations. Soon, others will copy this new information/perspective, and the guide can search for nonpareil content to add to their basket and keep it fresh. The bottom line is that a tour guide should have something to tell that others do not. While one can earn some expertise in art, another may specialise in cuisine, yet another may focus on the social dimension. But it is not just about the information you disseminate – it also about how you deliver it.

Humour

Humour is the best tonic to rejuvenate a sulking tourist group. "Look at the beautiful gold crown of the deity and the long necklace. People are poor because gods have all the gold!". Use humour selectively and keep the atmosphere light-hearted. "It's difficult for me to handle one wife; this king had seven".

To summarise, tour guides may consider the following to make their narrations more entertaining:

- Try to make use of all five senses of the audience – sight, sound, smell, touch, and taste.
- Use visual, audio/ music, props, interaction to augment the effectiveness of narration, and facilitate experiential learning.
- Remain calm and offer guests a friendly, relaxed, informal company. Do not stress themselves or guests. An informal approach rather than a formal tone of the narration is always advised.
- Establish cause and effect relationships to explain. Tour guides must try to correlate. "Hindus do not eat beef as cows are considered holy." "People put a sandalwood *tilak* on their forehead as the forehead is considered sensitive and tilak serves as a protective film against the unwanted rays."
- Tour guides may selectively exaggerate size, time scales, and other things to add spice to commentary. For example, "The roof of the dome that supports this huge chandelier can endure six elephants at the same time".

Chiselling themes

Consider the following in shaping the themes:

- Make themes personal by using words like 'you'. For example, instead of saying, "This monument is built in Indo-Islamic architectural style", a tour guide may say, "You need to appreciate the outcome when two great cultures come together and create a confluence like Indo-Islamic architecture, an example of which stands in front of you."
- Connect themes with things of symbolic importance. For example, instead of, "Cuisines have a history", a tour guide may say, "Cuisines have a cultural context and evolve over the time with the community."
- Use an analogy to explain: For example, instead of, "The breeze cools as it passes through these grills", a tour guide may say, "As the breeze passes through these intricate latticework, it suddenly expands as in the venturi effect making the whole area more pleasant during the hot summer."
- Use a metaphor that helps to connect, to explain or present a person or thing. For example, instead of, "Sarojini Naidu was a great poetess", a tour guide may say, "Sarojini Naidu was the Nightingale of India."

- As far as possible, use verbs in active voice. For example, instead of saying, "Taj Mahal was built by Shah Jahan", a tour guide may say, "Shah Jahan built Taj Mahal."

Techniques of interpretations

This section presents the nitty-gritty of interpretive communication for tour guides. It elaborates on the issues that are important to make narration effective and powerful.

Top visual priority (TVP)

Top visual priority is an essential concept in tour guiding. A tour guide must always look for top visual priorities. Talk about what people can see and engage their interest. Too often, guides launch into a lengthy historical background at places while something exciting is taking place. No one is listening!

There are two situations. The first time visitors areexposed to a landscape (moment of truth), they are mesmerised by specific characteristics which stand out or are strikingly different for them. They capture visitors' attention over other things. These are TVP tour guides must explain/interpret these first.

Sometimes when a tour guide is rendering a narration of some artefact/ monument, some non-routine things occur in the vicinity. For example, while you are telling about a building in the city, a milkman passes on a motorcycle with large cans of milk; or a loud hawker passes by, selling some local stuff – these capture the group's attention. This distraction has become a TVP. Take a break. These must be explained/interpreted first.

Sometimes there are too many elements in a tourist's gaze, for example, the skyline of a city. Again, the tour guide's responsibility is to help the visitors figure out what is worth seeing. In these cases, the guide should identify what should be the top visual priority for the visitors in a gaze.

Position triangle

One of the basic principles of compelling narration is to have a triangular arrangement with the object (being explained), guests and speaker positioned at the triangle's three vertices. This triangular formation allows the guest to clearly see the object without the guide obstructing their view as he/she tells them about the object facing the guests. On the other hand, the tour guide can look at the guests while speaking and occasionally glance at the object without disrupting communication.

Do not talk as you walk

One of the fundamental sins in group narration is to talk as you walk. If the group size is more significant than four, make it a practice not to talk as your walk. By doing so, the tour guide will stress their guests to run to catch up to be able to hear what is being said. Instead of enjoying the place, the guests would be busy competing to get close to the tour guide. Given the physical limitation of some guests, they might miss out on what you have told others while you were walking with them. They will feel deprived, and a perceived inequity will set in. They will gradually lose interest and may become recluses. The tour guide should always walk to a point and wait for the entire group to collect and be in a position to be able to listen to them comfortably. Once all group members are ready to listen, the tour guide may start speaking.

The 120° formation

Always ensure that the guests collect in front of the guide and stand facing them in a range of 120°. Ensure that the group members are not obstructing the view of other groups and visitors. They should stand in no more than two rows. The guide must be able to see everyone within the 120° angle of their two arms. Guides must ensure that no member of the group is standing behind them or even in line with them. A trick of the trade is to stand at a premeditated place and allow the group to crowd around them. Once the group has settled, the tour guide may subtly step backwards. The audience will now be in their 120° range.

4

Communicating time

Often the guide will have to communicate time to the group. For example, they may allow the group time to go around the site independently and return within an hour. Alternatively, they may require them to collect in the hotel lobby at a particular time in the morning. Instead of telling them to meet after one hour or by 4.00 pm at the main entrance or by 8.00 am in the hotel lobby, tour guides must peg time to their watch: "It is 3.25 by my watch. We will collect at the main entrance in one hour that is by 4.25 by my watch." Avoid vague instructions.

Gesturing numbers

It can be helpful to use gestures to create clarity in conveying numbers. For example, tour guides may use fingers to reinforce the numbers. One may say, "after you have visited the place, let us meet at gate number seven after one hour", and the guide may use fingers of his hand, also indicating seven.

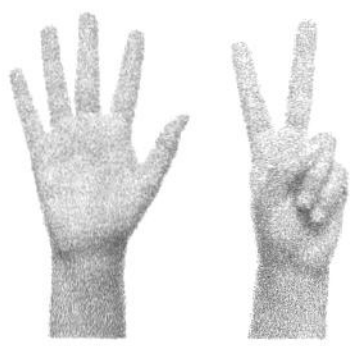

Use of technical terms and vernacular keywords

As you speak, you may use terms and words in vernacular which is difficult for the visitor to grasp. There is a two-step solution to this challenge. First, repeat the word slowly so that each syllable is distinctly audible. The second step is to provide a translation – word by word and conceptual (as a whole). It is always good to incorporate these words and let the tourists repeat them, which turns into a fun activity. However, never overdo it.

For example:

"Friends, this is Hawa Mahal... *ha-wa* .... *ma-hal*... '*Hawa*' means 'wind' and '*Mahal*' is the palace. So you can call it 'palace of winds' or 'the wind palace'. The building was so designed that the royal family could sit in windows and watch the busy street while enjoying the cool breeze."

Another example:

"This majestic gate In front of you is '*Haathi Pole*'...' *haa-thi*'...' *pole*' or the 'Elephant Gate'. '*Haathi*' is 'elephant', and '*pole*' means 'the big gate'. Kings would ride past this gate sitting on an elephant, so it is called the 'elephant gate'."

Eye contact

Establishing eye contact is fundamental to interpersonal contact. It means talking to someone individually by looking into their eyes. Eye contact builds a link with the audience. In case of a talk or discussion with a group, a tour guide must maintain eye contact with each member of the group:

- Stand facing the group, and ensure that they look at all the members – do not leave anyone out;
- Focus attention on the whole group;
- Even during the walking tours and site tours, face the group throughout the commentary. The tour guide must convey that they are interested in each one of the group members.
- Be aware of cultural differences; for example, guides should not impose eye contact on Japanese visitors who might not prefer it.

Often a tour guide will find themselves in a situation where they lead the group and talk to them in open, bright sunny outdoors. They are tempted to put on glasses, but this must be avoided. They may wear sunglasses while moving from one point to another. However, as and when they stop to talk or comment, they must take off the sunglasses – put them in their pockets or over their head so that there is no loss of eye contact with the group. A hat in such a situation can be used if necessary.

Personal territory

It is advised to keep a decent physical distance from the guests. We all have our personal territory – the space around us – but personal territories can vary both culturally and ethnically. Tour guides must understand these differences. As we saw in Chapter 3, the terrory has four zones, intimate, personal, social and public. A professional tour guide should usually use social space and sometimes personal space. They should never violate intimate space, which may be offensive to the guest. On the other hand, tour guides operating in public spaces may not be cognoscible.

Allow time

Tour guides should allow guests to venture on their own. They must be allowed time to internalise what they have seen and heard. Lengthy commentaries are despised. Short intermittent commentaries with ample time in between to explore are highly recommended. Also, guests, today want to take pictures. The time allowed for photography is also the time to reflect. So it is essential to recommend them good spots and allow them time for the same.

Briefing

Every time a tour guide accompanies a group to a new place, the group should be briefed about the essentials: the approximate time they all are going to spend there; the terrain of the site; whether tickets are to be purchased and who purchases them; the photography policy at the site; location of washrooms and restrooms; and other relevant things.

Iconography

Over time humans have used icons as part of communication to convey messages. Icons have transcended into local art and architecture and have become an integral part of the tangible culture. Communities have used these icons and symbols often for centuries, and while the use of symbols is preserved as part of the legacy, the meaning often gets blurred or even lost. Visitors are often inquisitive about what these symbols mean. Tour guides are expected to be versed in local iconography – the study of the symbolism of a work of art. A tour guide must interpret what a symbol, statue, painting, or architectural feature related to religion, art, entertainment, and other aspects of life convey.

For example, the symbol of the cross conveys Christianity, while the crescent symbolises Islam. Often religious tales and fables are depicted as paintings and murals. A tour guide must connect the visitor to the underlying story. For example, Virgin Mary is often depicted in a blue robe to convey heaven and spirituality. She is shown holding a vase with lilies and a bowl or urn of water, symbolising purity. Saints and prophets are often shown with a halo that connotes saintly wisdom.

It is forbidden to depict humans in Islam, so an Islamic monument will rather have floral motifs. Sikh architecture, influenced by Islamic styles, frequently uses the onion dome, frescoes, inlay work, and multi-foil arches. Ancient Indian temples extensively use icons and symbols to convey philosophical messages. Let us have a look at how a temple pillar from an ancient Hindu temple can be interpreted.

Figure 4.4: Interpreting the twelve elements pillar "*Dwadasha Lakshana Sthambha*" of "*Vishnu Kanta*" (Octagonal) class of Hindu temple pillars from Sun temple, Modhera, Gujarat in India.

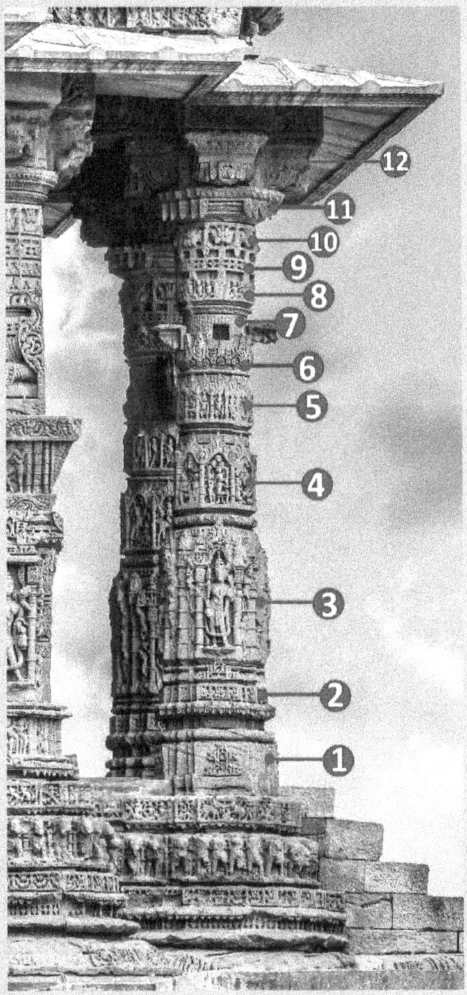

1 *Ghata/Ahvaana Patta:* Base water pot like *Kalasa* or Welcome Band generally contains *Sri Ganesha* or *Sri Maha Lakshmi*.

2 *Simhamukha Patta:* Lion Band – represents valour and courage.

3 *Dwarapala Patta:* Guardian Band – Guardians of the presiding deity in the temple are present.

4 *Vadaka/Vadaki Patta:* Musicians Band singing in reverence of the god.

5 *Avatara Patta:* Manifestation Band – different *avatars* of the presiding deity.

6 *Kalika Patta:* Lotus Bud Band – indicates the potentiality of everybody to attain moksha.

7 *Khandika/Ghanta Patta:* Bell band – signifies this universe's impermanence, transient creation through sound.

8 *Deva Gana Patta:* Embodiments of divine features – resembles all good characters of divinity.

9 *Gandharva Patta:* Geometrical Band where *Vajra* or *Padma Kulika* are present. *Vajra* indicates indestructibility. *Padma Kulika* indicates the probability of *moksha* all human beings.

10 *Kirtimukha Patta:* the embodiment of dedication to his master *Mahadev*.

11 *Kumbhika/Purnakumbha: Kalasa* – water pot at the top.

12 *Sirsa* Head/capital of a pillar – represents how important *Sirsa* is in bearing weight like the head takes care of a human body.

Ghata at the base is water, *Kumbhika* at the top is water, which indicates how life emerges in water and ends in water.

Credits: Photo and interpretation by Ramakrishna Kongalla, iconologist and academic

Key points

Visitors want to learn, see, and do! So they travel to sites for a mix of experiences. They travel to see, experience and learn (edutainment) about natural or cultural landscapes, sites, features, objects, people, events and stories.

Interpretation communicates what is noteworthy about places, people or events – why they are appreciated? It is a celebration of place and things, culture and nature, creativity and folklore, great successes and failures in history. Interpretation is a process of provoking audiences to think. The job of a tour guide is to engage the visitors through their storytelling so that they are hooked and make conscious efforts to construe the meaning.

A good interpretation makes a visitor inquisitive, motivates them to learn more, establishes the importance of the place, customises the experience and the story for the visitors, and finally, makes the tour guide's performance enjoyable, inspirational, impressive, and persuasive.

This chapter focuses on personal interpretation, where tour guides interpret a place or object during a physical face-to-face interaction. In contrast, nonpersonal interpretation is narrative depiction through exhibits, audio-visual aids, printed material, and other means without mediation by a person.

Sam Ham has proposed a TORE model for effective interpretation. He suggests that there needs to be an underlying theme (T) in the narration; commentary of a tour guide needs to be organised (O); interpretation needs to be relevant (R); and finally, the narration needs to be enjoyable (E) for the audience.

For an effective organisation of ideas, tour guides must avoid transferring too much information, keep their narrations simple, craft a persuasive elevator pitch to their interpretive narration, which may be structured using the"must tell, should tell, and could tell" model.

Tour guides must try to make interpretation meaningful by vividly using illustrations, drawing comparisons, using time hooks, building connections with what was earlier visited, and drawing parallels from the visitors' context.

We also discussed some key techniques of interpretation like the concepts of TVP, position triangle, not to talk as you walk, 120° formation, gesturing numbers, communicating time and vernacular terms, maintaining eye contact and remaining aware of trespassing the personal territory.

Interpretation is an obligation to provide uncompromising quality for the visitor's experience and always leave them asking for more! It is a love of talking to and with visitors and getting them as excited about the site as you are. Without this attitude about interpretive quality and customer care excellence, the site and visit become stale, boring and lacking soul.

Iconography is the study of identification, description, classification, and interpretation of symbols, themes, and subject matter in the visual arts. A tour guide must interpret what a symbol, statue, painting, or architectural feature related to religion, art, entertainment, and other aspects of life convey.

Key terms

Briefing	Short talk to pep up and share essential information and details. Generally at the start of the tour, day, and on entering an attraction, sometimes also at the end.
Commentary	A narrative used by tour guides to describe a gaze and provide information about various aspects of the place being visited by the tourists while on tour.
Elevator pitch	A brief, persuasive remark that guides use to spark interest in what visitors will see and experience. This typically explains why it is important, why the visitor should be interested in it, who it was for, what it did, and why it was needed.
Eye contact	Visual sense being dominant for most people, eye contact is vital during a tour guide's communication and conversation with their audience. Keeping eye contact with the audience conveys that guides are talking attentively to the guests and ensuring that guests actively listen.
Interpretation	Provides tourists new insights and understandings of the area they visit and the culture and environment they experience to make meaning and be more mindful. Interpretation is the essence of a tour guide's commentary or narration.
Links	Used to create an association between the object being observed and what guests have seen earlier on the trip. Links could be contrast or parallels.
Mental picture	A tool used by experienced tour guides who try to portray an imaginative scene, often historical, using words for its creative description. Painting a mental image requires visitors to 'imagine' the picture/ scene as the guide describes it in words.
Narration	The account of the place, product, attraction, happenings, related stories, facts and figures and other things rendered by a tour guide. Terms 'narration' and 'commentary' are used interchangeably in this chapter.
Nonpersonal interpretation	Refers to interpretation through media like signage, brochures, exhibits, self-guided trails to orient, and educate the visitors without a role for interpreters
Personal interpretation	Face-to-face interaction of tour guides with visitors. Guides decide the content and sequence of flow of information and even the routing. It is available in various formats - guided tours, walks, talks, demonstrations, and others.
Personal territory	The space surrounding a person they regard as psychologically theirs and feel comfortable within its bounds. Tour guides should not encroach on guests' personal territory, which could make them uncomfortable, angry, or anxious.
Theme	The main point or the central message in the interpretive narration developed by a tour guide for different audiences who visit the site. A theme summarises, and articulates the underlying idea that a tour guide wants the guest to perceive.
Time hook	A technique to connect the antiquity of the place or event with some event or monument from the visitor's context to relate with it and feel a sense of time.
TVP	Top visual priority is on top of visitors' attention over other things in a gaze. A tour guide should, on priority, talk about things that have captured visitors' attention.

Discussion questions

1 What is your understanding of interpretation? How is it different from traditional commentary by tour guides?

2 What constitutes a good interpretation? How do themes help shape interpretation?

3 How can nonpersonal interpretation tools help tour guides enrich experiences for their patrons?

4 Prepare a researched note on the use of innovative technologies in leveraging visitors experience at a site, for example, an archaeological site or a museum.

5 Elaborate on Sam Ham's TORE model to make interpretations more organised and relatable.

6 According to your experience as a tourist, what can a guide improve? List five points.

Activity

1 Visit a tourist attraction nearby. Research it. Develop two interpretations – one for school children and the other for baby boomers from Europe. In either case, develop an elevator pitch and an outline commentary using a must tell-should tell-could tell framework.

2 Visit a heritage site, read some of its histories and try to paint a mental picture at a location of your choosing.

3 Visit a local museum or monument, select an architectural element, and interpret it.

References

Colquhoun, F. (2005). *Interpretation Handbook & Standard.* Department of Conservation, Wellington, N.Z. Retrieved June 25, 2018, from: https://www.doc.govt.nz/documents/about-doc/role/policies-and-plans/interpretation-handbook-complete.pdf

Ham, S. H. (2013). *Interpretation: Making a Difference on Purpose.* Colorado: Fulcrum.

Tilden, F. (1957). *Interpreting Our Heritage.* North Carolina: University of North Carolina Press.

Urry, J. (1990). *The Tourist Gaze.* London: Sage.

Veverka, J. (n.d.). *Why heritage sites need interpretation.* Retrieved June 25, 2018, from John Veverka & Associates: http://www.heritageinterp.com

5 Planning

Learning objectives

After studying this chapter, you should be able to understand:

- The importance of planning for interpretation
- Research for a guided tour and sources of information
- The concept and importance of interpretation goals
- The issues in designing the interpretive plans
- How to develop the tours (products)
- Nuances of a tour guiding career

A tour guide typically appears to be a light-hearted, cheerful extrovert who engages with guests and maintains an upbeat tempo throughout the interaction. However, this high-spirited interaction has to be meticulously planned and executed. A tour guide's job is a professional performance with fine detailing. This chapter will discuss the prerequisites and the planning process for interpretation. What appears to be a spontaneous tour guide response usually has inputs from careful planning and rigorous homework. While a good interpretation is often considered an art, it requires methodological planning. Much research and planning go into designing interpretive themes and last-minute improvisations. Tour guides must be aware of the local resources – both natural and built. They must understand the psyche of their visitors, the travel motivations and the demographics of the key segments. They must also address the interests of other significant stakeholders – the community, the local government, and the company if they are representing one.

Effective interpretation is not by chance. It is thoroughly planned. The plan is developed around a few fundamental issues – What do the destination and each attraction offer visitors? What kind of people (age, background, nationality, etc.) visit these places and why? What do they want to know, or what do they want the guide to help them with? What are the constraints (time, money, effort, etc.)? Moreover, are the tour guides' meeting the guests' expectations?

Planning is the act of thinking about the course of action to achieve the desired goal. It is deciding beforehand what is to be done, when and how it must be done and by whom. It is about anticipating situations and preparing for them. This requires logical thinking and rational decision-making. Planning will begin with setting interpretation objectives and targets by assessing the resources at the destination, the limitations, and the potential demand for tourism products. Having a fair idea of what is to be achieved, the next step is to have a plan to attain the same in future. Planning is followed by implementing the plan and monitoring to ensure that interpretation is carried out as per plan and to the satisfaction of the patrons. Planning of an interpretive programme thus involves five steps: research, setting interpretation goals, designing the product, product development, and monitoring.

Figure 5.1: Planning the interpretive programme

Research

Any business is essentially matching demand and supply. Who demands and what is demanded? What can be offered? Thus, the design of any interpretation programme deals with two components: the destination resources that can be offered as products and the target audience for whom interpretation has to be designed and developed. The first step in any planning exercise is research. Research about the destination and attraction – the vital information about it; and research about the targeted segment of visitors to this place – who visits and what are they interested in.

Figure 5.2: Research at a destination

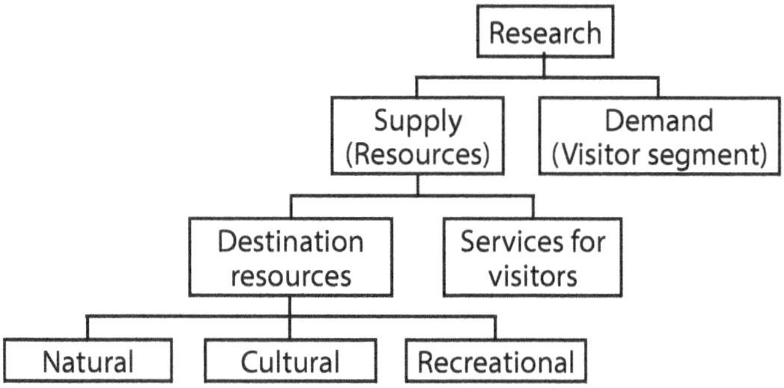

A tourism product is essentially about a visitor's experience based on visitors connecting with these resources, which in turn depends on how these resources are interpreted for the visitors. Therefore, the starting point is to identify the resources available at any destination. Resources are of two types: one, the attractions – natural, cultural, and recreational; and two, the services or available facilities that augment visitors' experience. Tour leaders must be aware of all such resources to bring them in for their customers. Thus, research at a destination must revolve around the dimensions identified in Figure 5.2 above.

Usually, an interpretation already exists for the popular attractions at a destination. There is a legacy of tourists visiting a particular place that establishes which visitor segments are attracted. The first step is to research the existing interpretative plans of the place. Subsequently, the tour guide can improvise the same. A guide is not expected to know everything about the destination, but their knowledge of local resources, products, references is often helpful for their patrons. Over time, tour guides have a fair understanding of the visitor segments that frequent the destination and what they usually look for and ask for.

■ Resource research

All the cultural, natural and recreational resources together contribute to a visitor's experience. The job of a tour guide is to connect a visitor to these resources through a well-designed interpretation plan. Resource research is carried out at two levels: first, to explore the significant destination resources to develop a theme for interpretation. Once the theme is identified, detailed research on this is required. One should look for the destination's positioning – which is why tourists primarily visit the place. Observe, talk to local tourism authorities, senior tour guides, destination management companies, and look for published literature about the destination. Second, one must understand which resources at the destination resonate with visitors. If the guide is unfamiliar with the local resources, they cannot connect with visitors. They must have the latest and precise resource information to exude credibility and curate visitors' experiences.

Attractions like archaeological sites, museums, parks, etc., often have a repository of information built over the years. Thus, an inventory of resources already exists for these. The authorities already have an interpretation master plan, creating content accordingly. Resource specialists continuously research and update information and add new resources to the inventory. The content developed by the authorities also helps tour guides understand the interpretive significance of the resources. Authorities generally invest in infrastructure to support their interpretation master plans. This development comes in handy for the tour guides. Remember, resources are silent until interpretation gives them a voice. So, let us now identify some of these resources at a destination.

Basic information and services for visitors

History of town/ city: the year of establishment and time hooks, political & business history, cultural legacy, speciality, if any.

Geography: the location on the globe and on the tourist map, flora and fauna, landscape, etc.

Access: by air, waterways, road, rail (nearest railway station), the distance from the important nearby cities, and details of rails, toll roads, skid roads, railroad rights-of-ways, bridges, historic highways, etc.

Accommodation facilities: number of hotels, type/ category of hotels, tariff plan, etc.

Other services: transfer services, car rentals, camping gear rental, availability of interpreters at different sites at the destination & at other sites.

Relevant services: hospitals, police stations, convenience shops, handicraft shops, etc.

Nearby tourist destinations: their importance, how to get there, etc. Always have a list of things ready for the tourists to do in their spare time.

5

Natural resources

Climate: seasons, average temperatures, rainfall and humidity, snowfall, sunset and sunrise, etc.

Type of habitat: forest, grassland, aquatic, desert, mountainous, polar, etc.

Flora: The native flora, garden flowers & weeds. History and significance of the flora, etc.

Fauna: Typology, important terrestrial and aquatic animals; vertebrate and invertebrate, their relation with the local community, etc.

Land usage: cropping patterns, irrigation, etc.

Seasonal features: Monsoons, tornados, cyclones, flower blooms, birds' migration, spawning, etc.

Geological characteristics: rock and soil types; earthquakes, volcanic eruptions; fossils, etc.

Notified area: parks, sanctuaries, bio reserves, and their significance.

Tour guides should be aware of local rare or endangered species. They must also be able to relate the destination's natural (geological and biological) and cultural resources.

Cultural resources

History: political & social history, historical events, etc. The legacy and connection with the present.

Historic structures: temples, palaces, forts, etc. Guides should be familiar with their timelines, time hooks, construction technology, etc.

Archaeology: archaeological sites, historicity, important finds, museums' artefacts, etc.

Cultural places: temples, museums, stadiums, traditional markets, historic sites, city centre, etc.

Cultural event: concerts, festivals, fairs,

Cultural landscape: architectural styles, traditional use of the area, landscaping,

People: of the place, their history, traditions, present status, etc.

Cultural elements: dance, music, cuisine, dresses, handicrafts, local produce, etc.

A tour guide must also connect cultural resources with natural resources.

Recreational resources

Key activities: hiking, biking, trekking, rafting, climbing, picnicking, snorkelling, bird-watching, sky gazing, dune bashing, off-roading, angling, swimming, bushcraft survival, or whatever the destination is famous for.

Places for primary activities: beaches, riverfronts, lakes, picnic spots, campgrounds, shopping districts, walking trails, fairgrounds, ranches, etc.

Visitor experience: what is the best time to visit a place or participate in an activity? What is a good view for photography? How can visitors engage with the host community? Things like traffic, crowd, pollution, etc.

Recreational add-ons: heritage walks, gastronomical experiences including indigenous cuisines, traditional bars, cultural centres, photo safaris, traditional games and storey tellers, traditional healers, curio and craft factories, etc.

Visitor data: visitor traffic, segments, demographics, seasonal use patterns, etc.

Tour guides must remain aware of any emerging recreational opportunities in future and their compatibility with the organic image of the destination.

Sources of information

An essential part of a tour guide's job is to make efforts to collect and collate information, often during the low-demand season. Quality of information is therefore important. Begin by scanning the information brochures published by the companies that engage their services and those who operate in the market. These companies have already invested time, effort and money in research.

At times, tour guides may have to operate at remote and less known sites. Especially in developing and underdeveloped places, information about these sites and attractions might exist but has not been digitised or made readily available over the Internet. However, some authentic and well-researched information may lay with the local library and government archives. Sometimes, enthusiastic journalists come up with featured articles about different aspects of the place, its life and tourist attractions in local newspapers, magazines, etc. This information can come in handy. Tour guides may also read published guidebooks like *Lonely Planet*, *Let's Go*, *Rough Guides*, and *Insight Guides*s.

It is easy to access information from the Internet, however, it is a plethora of information. Therefore, one may find it challenging to deal with the volumes and figure out if relevant information is updated. For example, view documentary films and video clips about the destination and its different aspects online, especially on YouTube. Likewise, a tour guide must remain updated with information. One way is to go to Google and set 'Alerts' with keywords. Then, you would receive emails whenever relevant new data is available on the Internet.

Much information is also available in the social domain. Especially in the oriental world, where information is not all documented, the community is a noteworthy repository of history and folklore. Therefore, a guide must try to source information from these non-conventional and scattered sources, but never fail to mention that the source of this information is folklore. Furthermore, they should participate in orientation programmes, tour management training programmes, and destination specialist programmes organised by the local tourism organisation. They should visit the local universities and colleges to meet experts in history, geography, sociology, anthropology, archaeology, etc. Similarly, they should interact with experts from various relevant authorities: archaeology, forest, zoological departments, etc.

An essential source of information is socialising with other colleagues in the trade. Tour guides may watch selected TV documentaries; radio programmes, and podcasts are also important. An interesting source of information are elders and knowledgeable people from the community who may provide interesting insights and anecdotes.

Finally, look for maps and undertake a recce of the destination. Tour guides must know their destination well. Research and collection of information are initially tedious, and gradually, with experience and practice, the tour guide needs to brush the information regularly.

Managing the content

Often the availability of information is not a challenge. However, managing and retrieving information is challenging. The first step is to lay a hand on the relevant material. Today, it is easier to organise data using smartphones to capture pictures and scan documents. Thus, guides can quickly build their repository: buildings, historical sites, architecture, cultural practices, flora, etc. Next, they must identify which resources are important and popular among visitors. Which is unique? They must also collect vital facts like important dates, names, measures, materials, etc. Another critical dimension is information regarding the fragility of resources. Finally, they must be aware of how visitor activity may impact the resources, and continually advise guests about the same.

The next step is to assimilate the information and regularly update notes with new information. If the tour guides' area of operations is broad, they may prepare subject-wise files, contact listings, and diaries. A tour guide must always have maps and an itinerary. Route maps can be easily prepared with the help of Google Maps or other apps. Finally, the tour guide must always visit sites where they will actually lead the guests. They must conduct practice guiding tours – calculate distances and timings, stops, artefacts and views at those stops, etc. They must prepare notes of key points/information at those points.

Data collection

Data collection typically involve three steps: *Capture* → *Record* → *Interpret*. First, the tour guide must visit the site to collect the information. They must visit the sites at different times of the day, week, month, and year to capture the different experiences. They must use technology, especially cameras, to capture their perspective of the place. They may also obtain photographs and video from other sources.

The next step is to record the information where they may classify, segregate and label the data. They must also note down the dates on which the data was collected and document the conditions under which it was created/ captured.

The third step is to interpret the data into a communicable message. Objectivity must be maintained, and it must be ensured that no facts are distorted or misrepresented. At this stage assistance of experts must be sought to interpret the natural or cultural resources, to verify the details and interpretations. An essential part of interpretation is to indicate the relevance of the resource. Thus, information about the impact must also be collated.

However, it must be borne in mind that both the sources may not be objective and may carry the bias of the source – the original author or the subsequent interpreter. Therefore, tour guides must validate the information obtained from various sources.

■ Establishing validity

It must be understood that not everything available from the above sources is reliable. Therefore, tour guides must make extra efforts to assess the quality and trustworthiness of the information, and continually review and update the information. Ensure that it is logical and defendable.

Tour guides should establish the validity of the information. It must be *triangulated*. That is, every piece of information should be verified from at least three sources, preferably primary sources. This exercise reduces the chances of misinformation and misinterpretation. However, sometimes only one source is available, and in such cases, it is good to quote the source. Also, do not take it for granted that information has been going on for around years in guiding and therefore, it must be true. Therefore, always check before you present it as a fact.

In today's world, new research findings keep flowing in. Tour guides must quote from the latest research and stay up-to-date. Because content is readily available, they must also remain aware of the source, which must be reliable and reputable. Often, the available content presents the author's perspective, resulting in multiple or even differing views about the same resource or event. A professional guide should avoid such biases as far as possible. They must allow space for differing perspectives in their interpretation.

In the same vein, information available on the Internet must not be relied upon where much unsupported and parochial content is available. While the Internet is undoubtedly a handy source, adequate caution must be exercised while choosing the content – the reliability and credibility of its source must be established. These days, most attractions have a dedicated website that shares well-researched information about it.

Finally, tour guides are advised to record helpful information in an organised way to be readily available when required. These notes form the basis for guides' interpretive programmes. However, these notes are only for reference, and tour guides should not make scripts – they will not remember these anyways!

Researching visitors and segments

As discussed at the beginning of this section, tour guides should have a reasonably good idea of who their customer is. Thus, the nature of demand must also be researched. The interpretive equation has two sides: the supply side – the resource, and the demand – the visitors who seek interpretation and guidance at the sites. For any interpretation effort to be effective, it must be customised to the audience. Therefore, a tour guide must try to find answers to the following:

- Who is coming?
- Why are they coming? What is their motivation?
- Which aspect of the site is important or exciting for them?
- Therefore, what should they know or remember about the resource?
- Moreover, how does visitor activity impact the resource, and what care must be taken?

Once the information about the resource is gathered, the next thing is to decide and design an interpretive message for the target audience. For example, suppose different audiences visit the site. In that case, there need to be different interpretative efforts to meet the expectations of each audience. Expectations and activities of visitors are often grounded in their backgrounds – circumstances, experiences, upbringing, education, etc. Therefore, to make sense of the background of their audiences, tour guides must understand their demographics. Research must consequently focus on collecting information about the gender, age, visiting profiles (if they are visiting as a family, adult groups with no children, single, etc.), source market (inbound or domestic, which places, etc.), level of education, their ethnocultural identity and experiences.

The next vital information to be collected is travel motivations, which is one of the most critical factors influencing the visit. It essentially asks the question "Why do people visit this place?" The answer to this question predicts visitors' travel patterns and buying behaviour. Therefore, travel motivation is essential in explaining the travel choice and behaviour, and is the key driver for all travel action. Many researchers have tried to explain the travel motivation of visitors (Dann, 1977; Crompton, 1979; Urry, 1990; McIntosh, Goeldner, & Ritchie, 1995).

A simple way to look at travel motivations is through *pull* and *push* factors suggested by Crompton (1979). Push factors for tourists are things like the desire for status, availability of vacation time and disposable income, workplace stress, etc. Understanding these travel motivators will help tour guides understand the expectations of their clients and serve them better. Pull factors provide reasons to choose a particular destination. Why does a destination pull visitors? Climate, attractions, events, etc. are examples of factors that attract visitors.

The success of tour guides depends on their ability to figure out the critical visit motivations and attend to them. These motivations could include escape, learning (about the place and people), socialising, having fun, searching for meaningfulness, self-esteem, etc. A tour guide will not use the same script for narration for visitors with different expectations. The tour guides should quickly reorganise the information and their thoughts and offer a commentary that captures the visitors' imagination helping them connect with the attraction. We have already discussed the *must tell-should tell-could tell* tool to organise the commentary in the previous chapter. Information about visitor motivations should be collected. For instance, if the group is young, they will be interested in taking pictures and knowing exciting facts.

Tour guides must also be aware of the usage patterns of the visiting segments – what they do at the site, where they go, and how much time they spend. This will help them plan the itinerary and also help them understand the impact of visitation at the site. If required, the tour guide should tailor the itinerary to mitigate the adverse impact of visitations – choose another trail instead of a busy one, or spend more time at some points than others to avoid congestion, etc. Thus information about the visitors' activity patterns should also be collected. This data can easily be collected from visitor centres, booking counters and portals, local transport, and parking lots. Study of usage patterns will help understand what attractions and services are demanded more (Veverka, 1994). The interpretative programmes must be designed accordingly.

Researching the significant others

Some critical stakeholders about whom the tour guides must collect information are the local community, the site authorities, travel intermediaries, tour operators, and travel agencies. The local community is generally both the product and the context. According to Ap and Wong (2001), a tour guide is a mediator and a cultural broker. As a mediator, they connect visitors to locals and their environment (de Kadt, 1979; Nettekoven, 1979; Pearce, 1982). Mediating is about leading visitors to their conclusions and facilitating them to learn. van den Berghe (1980), thus called a guide a middleman.

On the other hand, a tour guide is also a cultural broker (McKean, 1976). Cultural broking is the act of bridging, linking or mediating between groups or persons of differing cultural backgrounds to reduce conflict besides bringing about a change (Jezewski & Sotnik, 2001). Ap and Wong (2001) suggest that tour guides, through their knowledge and understanding of a destination's attractions and culture, and through their communication skills, transform visits from tours into experiences. Thus, tour guides must collect detailed information about aspects of the local community. They must identify critical cultural resources in the community and understand its perspective of the same. They must also understand the community's role and participation in guest engagement. However, always remember to be sensitive towards the host community while gathering the information. The other vital information that the tour guides must collect is about the leading authorities and decision-makers in different attractions and establishments. They must know who must be contacted when needed. Often there are visitor's issues that must be resolved.

As discussed earlier, a tour guide may use simpler, straightforward tools to gather data. In some countries and counties, secondary data is readily available. In other cases, the tour guide may collect through observations, informal interactions, or semi-structured interview protocols. They may also visit essential attractions and establishments to gather information.

- What is the general disposition of the community towards visitors?
- Who all are involved in serving visitors? Who are not? What are their primary concerns?
- What are their attitudes, beliefs, and values regarding relevant issues? Say, for example, now about hygiene and social distancing.
- Demographic and socio-cultural profile of the community
- Which are the organisations that are relevant for visitors and the trade? Who are the key people there?

Tour guides must also learn about the packages offered by the tour companies. These companies are often responsible for engaging the services of the tour guides. Therefore, tour guides may research the requirements of various companies. For example, they should know which company deals with which visitor segments and the time allocated for a particular attraction on their itineraries.

Interpretation goals

We introduced thematic interpretation in Chapter 4 during our discussion on Sam Ham's TORE framework. An updated discussion on this framework was presented in his book in 2013 (Ham, 2013). Both Tilden and later Ham believed that a relevant interpretation provokes the audience to think and elaborate. Thus, it is more about provocation than education.

The purpose of interpretation often is to present uninteresting and complicated technical information to leisurely visitors in a way that captures their attention and keeps them engaged for a while in natural and cultural settings. Therefore, thematic interpretation must be carefully planned and executed. Planners may begin with identifying the USP of the resource, then developing appropriate interpretive themes. To begin with, one must also differentiate between the subject and the theme. The subject is a broad issue or the topic, whereas the theme is a specific message.

People do not remember isolated, unrelated facts, more so, about a new place or thing from a different context. However, they tend to remember for long the underlying theme, which fades away only slowly. Thus, to create a lasting connection between the visitors and the visited place, interpretive naturalists, guides at heritage sites, docents in museums, park rangers and others develop interpretive programmes around carefully crafted themes. These learned themes, in turn, affects a person's attitudes and behaviour. Thus, having an exciting theme to talk about will ensure that audience is attracted to what the tour guide is telling and remains engaged.

Therefore, planners must identify the interpretation goals to find a relevant subject, the possible themes for interpretation, and then the sub-themes if required. In this section, we will discuss how themes and messages are developed. In the next section on designing, we will discuss that different themes will resonate with different audiences.

Themes always carry a message, the absence of which renders a narration unorganised and directionless. The audience finds it easy to follow when a clear theme is articulated. Once a theme is identified, the rest of the narration is woven around it. Having a theme makes it easy to select what to include and what to exclude – thus, organising the available information gets easy.

While subjects are single words or phrases, a theme is a complete declarative sentence with a subject and a verb and ends with a full stop. Themes are the underlying message of an interpretive exercise that conveys the big picture where the audience finds answers to questions like "so what?" or "what the big deal?" when a tour guide presents a place or event of a bite of history or contemporary life.

Furthermore, there could be several themes for the same subject. For example, the subject could be the history of the Chittorgarh Fort. A tour guide can choose one or more themes to talk about- "the fort is dotted with architectural marvels", or "the fort was quite invincible", or "the brave queens refused to surrender during the III siege."

Placing the subject and theme

Identify more than one theme for a subject by completing the following sentence:

"Talking about ___*(subject)*___, it is really very important for you, the audience, to understand that ___*(theme)*___."

Once a theme is identified, use R (Relevant) and E (Enjoyable) elements to fine-tune it. This polishing of narration has been discussed in Chapter 4. Once a theme is identified, up to four sub-themes may be identified.

Designing

Designing is essentially about matching messaging with the audience. Many times, tour guides burden themselves with gathering and disseminating information. Whereas, today, it is futile to compete on the information. Visitors are equipped with the gadgets to access information on demand. They do not seek basic information from tour guides. In most cases, they seek experiences. Tour guides facilitate immersion in experiences. (Ham, 2013) famously pointed out

> "Interpreters and guides usually worry too much about content. You can look that stuff up in a book or talk to someone who knows more about it than you. What really separates accomplished interpreters and guides from the rest of the pack is their ability to get inside their audiences' heads and leave something worthwhile there. It is their ability to communicate in compelling ways."

The job of a professional tour guide is to connect the audience with the resource by associating meaning to it. However, to make interpretation effective, the tour guide requires significant effort to attune the message to the audience. The audience can broadly be classified into captive and non-captive. A *captive* audience is one visiting the place for an obligation, like an official visit, a study tour or research. Such an audience is driven by external rewards and therefore have focused attention. They visit the place with more academic overtones. Keeping this audience absorbed and engrossed with a didactic interpretation programme is relatively easy.

The tour guide's challenge is to handle the *non-captive* audiences that are more often than not varied. These guests visit the place primarily for enjoyment during leisure. Their prime motive is recreation, and they have no obligation to learn and therefore remain attentive. With a shorter attention span, they tend to be distracted. Thus, tour guides need to design and develop interpretation plans for different segments of the non-captive audience likely to visit the place.

With advanced knowledge and understanding of visitor segments at a site as identified through research, the tour guides must attempt to match their messaging with the targeted segment of visitors. A good starting point is profiling visitors in terms of group size, age, place of origin, cultural, ethnic, racial and religious background, and travel motives.

Knowing the age group of the guests would help plan the appropriate route and length of the itinerary. For example, with elderly groups, a shorter route that does not have frequent ascends and descends, with more stoppages for rest would be preferred. On the other hand, a younger group may enjoy an uneven terrain. In addition, kids may run around, so the route may be designed to avoid intersection vehicular traffic.

Children process information differently from adults (Farnham-Diggory,1972; Gray, 2008). Thus their participation in experiences and learning is different, and this needs to be considered while designing the interpretive programme. Children are also not one monolithic group.

Kids have limited exposure and, therefore, limited imagination. Their experience is based on their happenings in their surrounding environment. As they grow up, they look for meaning and predictability through repeatability – that is what repeats in their routines. It could be meals or festivals or something worth remembering. At around six to seven years of age, they start getting inquisitive. They want to try new things. Their cognition includes measurements and, therefore, comparisons. Thus, an interpretation plan can include activities that involve search and comparisons. They also begin to have a feel of time – yesterday, today, and tomorrow. Together, they have some understanding of past and present and can compare the two.

Further, children have a shorter attention span and might not be interested in lengthy commentaries and serious stuff. So the tour guide may use enchanting tales and engaging games to keep the audience enjoying and interested. In the case of cultural resources, the tour guides may introduce heroes who fought valiantly or did something worthwhile. Again, these serve as benchmarks for comparison. Older children develop the ability to relate what they see and hear from tour guides to their experiences, and such inputs and comparisons help address their curiosity. Older children can connect their learning in schools and other sources to the cultural and natural resources being visited. Interpretation programmes can include tools like demonstrating, sensating, juxtaposing, linking, and drawing parallels, among others, as discussed in Chapter 4. Teenagers are capable of adding meaning to what they see and hear. They look at the interpretation from the lens of their values and beliefs. Reflection to stimulus is an essential tool here. Thus, the plan for teenagers may be built around provocation, interaction and discussion rather than feeding them with information.

On the other hand, the baby boomers and silent generation visitors have vast lived experiences that can enrich the discussion around the resources. In addition, many of these visitors have witnessed history as it was created. Thus, they add value to the discourse.

A large group will be handled differently than a small group. Their group dynamics would be different. Taking around a larger group is trickier than accompanying a smaller group. Stopovers will have to be carefully chosen to accommodate a larger group. The pace of a larger group would generally be different. Techniques like position-triangle and 120° formation discussed in the previous chapter will have to be adapted according to the group size and layout of the site. For example, in the case of a smaller group, a tour guide may do some talking while walking, which is not advised when accompanying a larger group. It might also require more time and effort to involve a large group in certain activities. Even supplies like pamphlets, water bottles, goodies, etc. would depend upon the size of the group.

Cultural background is a key determinant of the interpretation plan. Visitors from different cultural backgrounds must be attended to, to match their subconscious mannerisms and expectations differently. In Chapter 3, we discussed the need for cross-cultural sensitivity in tour guides. The use of space and non-verbal communication is of particular concern as they are likely to be misconstrued. Avoid any political statements and never take sides in a controversial topic.

Another critical dimension to consider and include in the interpretive plan is accessibility. Accessibility here has two connotations. First, the interpretive programme must be inclusive to include a wider section of audiences. For example, separate programmes may be designed for visitors with different needs. For this, the tour guide must be aware of his audience's physical and mental abilities. Second, more providers today recognise the need for multi-sensory experiences and, therefore, access through multiple senses besides just relying on sight – touch, smell, taste, and sound must be included in an interpretive programme. Tourist destinations are increasingly using technology to leverage interpretation programmes. In this case, the tour guide's role expands to encourage the visitor to use such technology and educate them for such use.

Based on their research of market segments, travel companies that hire the services of a tour guide design the itineraries. Interpretive programmes of the guides must adjust for available time, time of the day that the visitors are at the site and other variables like crowding. At the same time, the interpretive programmes must be flexible enough to accommodate some last-minute exigencies.

Product development

Product development is tactical. Having identified the targeted visitor segments and the USP of the resource, themes are developed. These must be translated into a route around the site and accompanying narrations. Tour planning includes addressing practical concerns and preparing for them. It also includes the rollout. This is about readiness – willingness backed by a tour guide's ability to execute a successful tour. Product development thus includes developing the exact itinerary, which has three dimensions: routing, scheduling and loading.

Routing is the process of creating the most cost-effective route through the minimisation of distance or travel time necessary to reach a set of stopovers in a tour. It lays out the flow of the tour at a site. *Scheduling* is the process of assigning arrival and stayover time at each stop – when will the group arrive at a place, and how long will it stay there. The fundamental objective of routing and scheduling is to cut down on physical effort, allowing maximum exposure to the place being visited. Thus, routing is like a railway map, whereas scheduling is like a railway timetable. Once the schedule is ready, the activities are loaded. So, *loading* spells out: what message must be delivered, what activities to engage the visitor, and how much time they will spend at these locations. However, tour guiding is not perfect math. At times following a fixed schedule is difficult, and it is okay to be off by a few minutes here and there.

Let us discuss some finer points that help tour guides lead the group more effectively. Professional guides leave little to chance – they anticipate the challenges. First and foremost, the tour guide must be organised in thoughts and deeds. Today, technology comes in handy. Many apps can help tour guides organise and schedule their assignments. They can keep a record of groups that have been assigned, along with the details of the contact person (email, phone number, home address and where they are staying), date of visit, place of the visit, and time of meeting among others. Apps also allow you to set reminders, maybe a week earlier. Contact with the group or its leaders is an opportunity to build a rapport with the group. There are two underlying benefits: one, the confidence of your guests would grow as they know someone beforehand at an unknown destination; two, when you meet the guests for the first time, you will be speaking to someone you knew rather than struggling to establish a rapport with unfamiliar people. Should they decide to cancel their trip, they should not forget to cancel you. Your reminders are a reminder that they have engaged you. It is always advised to establish contact a day before.

The second most important preparation component is learning a bit about the guests. While you already have researched information about the segment, initial contact is an opportunity to talk to your guests about their food preferences, their hobbies, their background (country, age band, religiosity, and work), their likes and dislikes. This initial contact returns two benefits. First, the more you know about them, the better placed you are to organise a trip and suggest some local attractions ahead of time. You may discuss this with them and plan an itinerary with them. The other advantage is that this serves as a pep talk conveying your interest (and care) in them! Please take note of everything and do not leave it in your memory.

It is advised that the tour guides carry a small bag or backpack, with all the necessary things. We discuss a tour guide's tool kit in the next chapter. This kit will come in handy in many situations. Remember to keep it light enough not to bother you but full of essentials you and your guests might need.

Plan details and rehearse

Even if you are an accomplished tour guide, it never hurts to practice tours and the narrations and improve them. Practising is particularly important when you are a beginner. Novice tour guides may identify different city tours that they may be comfortable organising. For each tour, list the places of interest and things to see (and be shown). Then sit down with a map and plan out the route in detail. Sometimes printed maps are not available. Use Google Maps. Also, use the 'Add destination' facility to map out your route. Balloons will show your stoppages. Take a printout. Other route planner apps are also available.

When you are planning the route and itinerary, consider the appearance of your city on that route. Plan your travel routes along the most scenic and beautiful roads, even if it takes a bit longer. A tourist will be seeing your city for the first time, and you would like your city to make a great impression. Take them past the signature buildings of the city. Do not force your guest through the ways and places that are an eyesore.

The next step is to write out detailed descriptions of the things you would be showing (ballooned stoppages) including, history, dates (minimal), and interesting little facts that only local people would know. You can create a slide show for your ready reference. Bulleted information about each attraction can be used as you travel between points. Especially if you travel on a coach, you will have the opportunity to glance through your hand-held device. You may request a friend, a family member or a fellow intern (trainee tour guide like yourself) to tour with you as a guest. Note the time taken during the various tour sections and make notes of the costs for food, snacks and transportation. Your mock guest may provide you with helpful feedback.

Once you have prepared your tours and practised them, it is advised that you memorise them almost perfectly. This cramming will help you concentrate on the needs of your guests while you are on tour, least you should struggle with the content of your commentary and visibly lose confidence.

Consider adding little extras to the tours. For example, you could organise a short demo of cheese making en route. Talk beforehand to the dairy owner. Tell them that you would bring guests for the demo. They may buy some cheese as well. Little extras make the difference between a good tour and a better one.

Prepare to introduce yourself

The first impression has a lasting impact. Everything related to the tour can be affected by a not-so-good first impression. When a guide has sufficient time to prepare, there should be no excuse for a poor start that will lead to misgivings on the rest of the tour. Instead, tour guides should leave a signature during the first interface. We often listen to radio jockeys, who each have a signature introduction. A tour guide may not be as dramatic as an RJ, but should design and practice a signature introductory para. Guides must plan for an impactful introduction.

Do not forget to include the essential information. Tour guides, to establish a rapport with the group, must start with greetings and introducing their name. Sometimes the name in vernacular is challenging for the guests to pronounce. In that case, it is advised that guides have a shorter and more straightforward nick-name for a routine conversation. Keep the information about yourself minimal and interesting. The introduction should invariably include the company that the tour guide represents. They also represent the place and welcome visitors to the place or site or the city. The next element is a brief overview of the tour, which is essential to set the expectations and sometimes moderate the group's expectations. Should there be a pressing concern related to the site being visited, it should also be highlighted right at the beginning. Finally, do mention any prerequisites for the tour, for instance, that footwear or camera is not allowed inside the temple or take care of your belonging in the crowded areas.

Here are some examples of commentaries that you can try

Example 1

On the first day, the guide meets the guests at the hotel lobby as they check in.

'Good morning, everyone! I'm Vikramjeet Singh, and I'm your host for the next five days. You can call me "Vicky". I can promise that you are going to have a great time in India. You can leave all your worries for me and enjoy your stay in India. We will begin here in Delhi. It is my city, and it is called the city of people with 'big hearts', and I am going to show you exactly why! After you settle and take some rest, we'll go around this

part of Old Delhi. Delhi is not only about politics, Red Fort and Lutyens' architectural marvels. It is about its marvellous traditions enshrined in its antiquity that dates back to the 6th century BC. Even in its present avatar, *it is around 400 years old, built as Shahjahanabad by Mughal emperor Shah Jahan.*

We will go around the old quarter of the city for around an hour. We will walk close to a kilometre – are you ready?

Example 2

The tour guide meets the group one of the mornings to take them around.

'Good morning, everyone! I'm glad you've joined me for this morning's tour in this lovely city of Nairobi. I am Biryomumaisho Mwakugu, but you can call me BM. Nairobi is very different in the morning. Nairobi is from the Maasai phrase 'Enkare Nairobi', *which means "cool water"....the river Nairobi which is right in front of you. I am going to show you what I am sure you'd miss on your own. This time it's not about dates... it's going to be about life...how we live in Nairobi....the green city in the sun.*

History is not about kings and queens only. It's about people as well...who lived here. I'll spend just a few minutes, first telling you about the people who built it and lived here in the past, and then we'll go and see what they've left for us to see and admire.'

Briefings

Do not take your guest by surprise. Allow them to be mentally prepared for the trip. If they have extraordinarily high expectations, these must be moderated. The first meeting should allow visitors an overall idea of the tour. Then there has to be a briefing for each day. Moreover, the tour guide should brief the visitors at each attraction or place being visited. Ordinarily, the tour manager is the one who accompanies the group for the entire tour and tour guides are engaged only for the day or for the specific site. Since the itineraries are planned, the points where the guests need to be briefed must be identified. Briefings should be planned and rehearsed.

The briefing should typically include details of the time that the tour will take, the distance that will be covered and the kind of terrain to be traversed, the weather forecast, the kind of preparation guests need to make in terms of clothing, footwear, carrying water, among others. In addition, each site has its camera or photography policy. Guests should be briefed about it. Furthermore, tour guides must never forget to share information about the availability of restrooms and eateries at the site. Finally, it is vital to share how one can get back to the meeting point or the hotel if one is stranded or left behind.

Besides making them at ease, the briefing will help them plan and organise themselves. Guests may identify any challenges they foresee and discuss them with the tour guide. Keep the briefing thorough, simple and easy to follow.

Tourist concerns

Taking good care of the tourists is a prerequisite of a good tour and an expectation from a professional tour guide. They must empathise with the guests and anticipate the challenge they may face. Rest, restrooms and eateries are three concerns for tourists on a trip. Once a tour guide interacts with the group, they may estimate the requirement for rest. How long can the guests comfortably walk before they may need to take some rest? They should plan the opportune time and stop on the proposed route for a rest break. Consider if there is enough and decent space for the group to rest. Is there a place with shade? Are there any eateries around, if required? Is there a restroom facility? If the group requires buying batteries or replenishing water or energy drinks, are there some convenience stores? Anticipate visitor concerns and plan for the same.

Plan the groups' stops

Plan for stoppages. As you gather a group at a point for a short comment, ensure that they stand conveniently, are not obstructing other visitors, and are not disturbed by passers-by, traffic, and people. Ensure that visitors do not have to stand in the scorching sun or an inconvenient posture for too long. Ensure that the group is not standing in the middle of a pathway or are blocking other guests' movement for a good view for long. All of this should be planned.

Often tour guides have a fixed itinerary for a site with predetermined stops. Suppose, at a monument, it takes 75 minutes to go around and eight stops where a tour guide delivers the commentary about a gaze. However, the tour company plans that the group on tour can spend only 60 minutes at this monument during the day tour. The tour guide is expected to adapt to this requirement. Often, tour guides follow the standard itinerary but rush through it, severely impacting the quality of experience for the visitors. Instead, the tour guide should plan a new local itinerary for the site with, say, only five stops that could be completed in 60 minutes.

Always be prepared that the unexpected might happen. Prepare for a contingency plan too. Never panic in a situation you have not prepared for, and instead, take it as a challenge and improvise.

Photography

Instagramability is one of the crucial factors in holidaying today. An emerging trend among the millennial cohort is photographing and posting on social media. Other visitors also want to cherish memories of an exciting trip for a long time. Photographs are the most tangible evidence of these memories, and this visit might be their last time in the city or at the destination. This visit might also be their first time to the place. A tour guide is better placed to advise photography – the points and the angles that capture a beautiful view. A guide must plan for these photography opportunities and acquire photography skills as they may have to help the clients with the shoot. Allow sufficient time for photography and include stops for photography in the itinerary. Advise the popular viewpoints.

■ Closing the tour

The closing of the tour must be planned. The last thing that the tour guides must do is let the tour close on its own, with energy and enthusiasm gradually fading away. Instead, there should be a definitive high end to the tour. The end-of-the-trip should be an opportunity to re-energise and re-motivate visitors. Remind guests of the trip's high points and make them feel that they have had a great visit. If it is the end-of-the-day briefing, reassure them that the next day will be exciting. Also, guests should not be pressured for tips. If they want, they will tip on their own. Finally, remember to end the day with three things – summarisation, directions for next steps, and energetic sign-off.

Summarise

Summarise the day, highlight what had been achieved during the day and try to connect it to the overall plan for a fun holiday.

Directions for next steps

Leave your guests excited with a fair idea of what they would be doing the next day. Give instructions for preparations for the next day's activities, if any. For example, tell them precisely when and where to meet. What clothes and necessities should they carry? Any other hints at preparation should come now.

Energetic sign-off

The trip or the day should not end on a low note. Instead, the tour guide should sign off in style with a premeditated goodbye note.

Monitoring

The last part of planning an interpretive programme is monitoring. This includes feedback and evaluation.

As visitors arrive at the destination, the tour guides interact with them. This interaction is an essential source of feedback. Often the travel companies that engage the services of tour guides have a formal mechanism of collecting feedback from guests. In addition, DMOs and site authorities also conduct visitor satisfaction surveys. These are all sources of helpful information to understand if the performances of the tour guides are in line with the expectations of the visitors. It also helps identify which dimensions and components of the interpretations are not resonating well with the targeted audiences.

Therefore, it is desirable that monitoring and evaluation be made an integral part of the planning of the interpretative programme itself.

What should be evaluated?

There are three critical elements in a visitor interpretative programme – the tour guide, the visitor and the program itself. For effective interpretation and commentary, the tour guides must be good enough to captivate audiences with their communication and interpersonal skills. A tour guide's performance is reflected in visitors' responses, which are based on meeting their expectations of visiting the place. The interpretive programme must be well researched and designed to address the visitor expectations. Thus, the deliverers must be evaluated around their verbal and nonverbal communication skills and techniques and the quality of their interaction with the visitors. The parameters to evaluate these include the quality of their content, grooming, body language, voice quality and grammar, enthusiasm, confidence and credibility, entertainment quotient, and engagement. How were the queries handled? How were the demanding guests handled? Did the guide display empathy and concern? Was he/she helpful?

The customer is the king. Thus, the superordinate objective of any interpretive programme is to return a satisfied customer. Therefore, customer evaluations should primarily assess overall satisfaction, and this must be evaluated along conceptual, emotional and attitudinal dimensions. The conceptual dimension is about what people learn from their visit – information, retention of interpretive themes, attendance and attention they pay to the tour guides' narrations. The emotional dimension is about what visitors feel, mainly about the enjoyment and enthusiasm during the visit. Finally, the third dimension is attitudinal, which is reflected in behaviour. The change in visitors' attitudes and behaviour may be monitored, including their participation, willingness to participate in the co-creation of experiences, and responsible conduct at the site.

Lastly, the interpretative programme must be evaluated. Was it provoking besides being relevant and factually correct? How was the content organised and delivered? Did the intended themes figure prominently and were retained by the visitors? Was the trip planned and executed correctly in terms of length, pace, stoppages, effort on the part of visitors? Did the briefings at the beginning and the end of the tour include all necessary information?

Who should evaluate?

Visitor feedback is the most direct and helpful way of assessing the effectiveness of the interpretation. Visitor evaluations help us with different aspects of the interpretation plan, including the interpreter – the tour guide. A tour is generally designed some time before. Evaluations help us understand the changes that have taken place and whether tour guides have accommodated them in their interpretive narrations. Likewise, visitor expectations change over time, and evaluations help capture those preferences. For example, some points, displays or exhibits have lost their sheen over time and need to be excluded from itineraries or re-packaged as part of another theme or sub-theme. Accordingly, tour guides and authorities receive inputs for the upgrade of themes or developing new themes.

Evaluations also offer feedback on a tour guide's performance during the tour. Travel agencies and tour operators who hire these guides and the guides themselves are interested in these evaluations. In addition, these evaluations are often reflected in ratings of the tour guides on trade portals or destination websites. As such, guides can also monetise their ratings by charging a higher fee for a better customer experience or as curators of specific offbeat themes.

Monument/ park/ museum authorities are responsible for the site. Sites may change over time – there are new exhibits, new venues, new themes and new services. Have these been included in the interpretive narrations of the guide? These agencies may evaluate the interpretation and provide helpful feedback.

Similarly, destination leadership are another substantial stakeholder for evaluation and feedback. In some cases, there are well-designed programmes for assessing changes in visitor expectations, evaluating the competitive requirement, and inclusion of politico-legal mandates, among others. They assume the responsibility of tour guide reorientation and retraining. In addition, they continuously evaluate guides and interpreters to identify gaps for training and improvement. Finally, they may evaluate for quality control and assurance purposes.

Thus, a 360° evaluation is possible for tour guides. Evaluators may include visitors, authorities, destination leadership, fellow guides, tour managers, travel agencies and tour operators.

Planning a career as a tour guide

Another critical issue that needs to be discussed is to plan a career as a tour guide. In the first place, you should evaluate whether you are interested in becoming a tour guide. Chapter 1 discussed the job description and the job specification for a tour guide. Given those inputs, would one want to become a tour guide? The following analysis is proposed.

The analysis begins by asking yourself if you want to be a tour guide. A tour guide's job might be fascinating, but would you like to work for this profile. First, analyse. Consider your personality – your likes and dislikes, your skills, among other things. A tour guide must be intrinsically interested in people –

extroverted and skilled in interpersonal relationships. A tour guide must know the place; be interested in the area's history, geography, and politics; excited about showing around your community and its assets to visitors; talking and taking unknown people on trips – helping them, taking good care of them. It is also vital to understand yourmotivation to be a tour guide. Will the factors that motivate you today sustain over time.

The next thing is to observe other tour guides. Observe them closely to see their routines, pressures and obligations. Also, talk to other people in the trade. Talk to experts and experienced persons. Talk to travel agents and tour operators to understand their requirements. Finally, see if all this discussion still suggests that you should become a tour guide.

It is also essential to figure out what would one like to do. A tour leader is many things. It can assume different shapes – leading on short domestic tours, or being with the group on longer trips. You must decide whether you would like to be an inbound tour leader handling international tourists to familiar places or handling outbound tours to a specific destination or if you would like to show around the domestic guests. You may be interested in a particular type of tour or with a particular segment of clients. Decide what interests you and find out if opportunities exist in the area you intend to serve as a tour guide.

Having analysed the market, your interest and profile, if you want to be a tour guide, you must start preparing for this role. Talk to tour operators who bring guests and the authorities of attractions about the existing interpretation themes. You may like to figure out the gaps. See what information inputs are required in history, geography, botany, archaeology, etc. Understand what kind of tourists you would handle. Work out what kind of skill training would you require. Decide the scope of your work – inbound or outbound tour leader, or a city or a site guide.

The next step would be to enrol with a reputed or recognised training institute for tour guides anr participate in the programme. Courses typically run for 2 to 3 weeks. More extended or formal educational programmes lead to academic degrees in some countries. In many developing economies, the tour guide training is supported by the local authorities who organise it when they determine such training is required. An aspirant may consult the local print media or website of specific programs to learn about these training courses. Guide training is generally organised on a full-time basis. However, some part-time options are available. During COVID times, many online training options are available. For example, the Ministry of Tourism, Government of India had launched an online programme, Incredible India Tourist Facilitator (IITF) Certification Programme (www.iitf.gov.in).

In some cases, tour operators recruit individuals for their interpersonal skills and interests. They subsequently learn on the job. However, given the state of competition, it is advised that one gets trained before getting into the profession. Training will also allow certification. In some countries, tour guides must obtain a professional license to work. Such license is available after completing specific requirements and acquiring certain skills. If mandated, the aspirant should obtain the necessary license after meeting the statutory requirements and completing the laid procedure. After that, the novice tour guide may register with the local tour leader/ escorts/ tour guide associations, if any.

Tour guides typically operate as freelancers. The novice must develop contacts with tour companies that may take them as an apprentice for some time before they are allowed to handle the tour party independently. Some companies may hire beginners to work as tour escorts under full or limited supervision of a tour manager. However, that is just the beginning. Rookie tour guides must embark on an unending journey to keep learning about their area, tourism, new products, and markets. Continuously learn about the changing tastes and preferences of their markets. They must enter the profession with an open mind, remain watchful, and one will learn every day.

Remain aware

- A tour guide's job is vulnerable to seasonality. There will be peak seasons when there will be much work, but you may struggle getting guests during off-peak seasons.
- It is not a routine 9 to 5 job, and may demand long working days. From accompanying tourists on an early morning city walk to ensuring a late-night light and sound show and seeing them back to their hotel. So, make sure you are capable of working a demanding schedule.
- It is not about enjoying travel. It is about helping others to travel and enjoy. While others enjoy, you will be working out the logistics for the next part of the tour.
- Foreign language skills are helpful but not mandatory. However, it is always advisable to learn another language.
- Should you have other skills, you can offer a speciality – leading culinary tours, art history trips and more. Adventure tours are becoming more popular,though to be hired as an adventure tour leader, you need to be good at outdoor activities.
- The job of a tour guide per se may not fetch much money. Interestingly, as discussed earlier, there are multiple revenue streams of commissions (networking is the key) and gratuities. Non-monetary rewards and satisfaction also supplement.

Key points

Good interpretive narration is not by chance. It is thoroughly planned. The plan is developed around a few fundamental issues: What do the destination and each attraction offer visitors? What kind of people visit these places and why? What do they want to know, or what do they want the guide to help them with? What are the constraints? Do the tour guides meet the guests' expectations? Planning an interpretive programme involves five steps: research, setting interpretation goals, designing the product, product development, and monitoring.

The first step in any planning exercise is research. Research the destination and attraction – the vital information about it; and research the targeted segment of visitors – who visits and what are they interested in. Part of the research focuses on resources at the destination and the tourist segments visiting it. Resources are classified into attractions and services available to visitors. Attractions are divided into cultural, natural and recreational resources.

Interpretation commentaries will be built on researched information. Quality of information is therefore essential. Guides may rely on primary and secondary sources. Often the availability of information is not a challenge. However, managing and retrieving information is challenging, and not everything available with the above sources is reliable. Therefore, tour guides must make extra efforts to weigh the quality and trustworthiness of the sources and information.

Tour guides should have a reasonably good idea of who their customer is. The nature of demand must also be researched. The interpretive equation has two sides: the supply side – the resource, and the demand side – the visitors who seek interpretation and guidance at the sites. For an interpretive plan, interpretive goals must be set. Then, thematic interpretation must be carefully planned and executed. Planners may begin with identifying the USP of the resource as identified in the previous stage of research. After that, appropriate interpretive themes need to be developed.

Designing is essentially about matching messaging with the audience. The job of a professional tour guide is to connect the audience with the resource by associating meaning to the resource. The tour guides must attempt to match their messaging with the targeted segment of visitors. As discussed earlier, profiling visitors in terms of group size, age, place of origin, cultural, ethnic, racial and religious background, and travel motives is a good starting point.

Product development is tactical. Having identified the targeted visitor segments and the USP of the resource, themes are developed. These must be translated into a tour of the site and accompanying narrations. Tour planning includes addressing practical concerns and preparing for them. The itineraries have three dimensions: routing, scheduling and loading.

It is advised that tour guides, including experienced ones, plan tour details and rehearse interpretive narration. In addition, guides may prepare to begin the interaction with the guests with a practised introduction of the self. Briefings are important. They are generally at the start of the tour, day, and as one enters an attraction. There is a briefing at the end also. Tour guides must also remain aware of tourist concerns, plan the group's stops, photography opportunities and points.

The last part of planning an interpretive programme is monitoring. Monitoring includes feedback and evaluation.

Another critical issue that needs to be discussed is to plan a career as a tour guide. In the first place, you should evaluate whether you are interested in becoming a tour guide. For this, you may observe others and talk to those involved in the travel trade. Aspirants must skill themselves by participating in training. They may also need to acquire certifications and licenses as mandated by local authorities.

Key terms

Briefings	Short talsk, which is there to pep up and share essential information and details. Generally at the start of the tour, day, and on entering an attraction. There is a briefing at the end also.
Cultural resources	Unique and nonrenewable, intangible (spiritual), and material phenomena (natural or made by humans) associated with human activities. These include sites, structures, and artefacts to which an individual or group attaches some value about its historical, archaeological, architectural, spiritual, and human (cultural) development.
Instagramability	The suitability of a destination (or product) as attractive or exciting enough for photographing and worthy of sharing on social media.
Interpretive goals	The communication goals to be achieved during an interpretive programme, including talks, activities, demonstrations, among other things, at a visitor site.
Introduction	A tour guide's first presentation to a group of visitors. It invariably includes a brief overview of the tour to set (and sometimes moderate) the group's expectations; welcome to the site on behalf of the tour company and the destination.
Loading	Spells out what message must be delivered, what activities to engage the visitor, and how much time they will spend at these locations
Monitoring	The systematic periodic evaluation of an interpretive programme's performance vis-à-vis its goals and objectives. It includes collecting, analysing and using the information to track the progress.
Natural resources	Resources like flora and fauna, landforms, rivers, beaches, lakes, sunlight and other naturally occurring features, attracting visitors.

Recreational resources	A broad range of outdoor and indoor public and private areas and facilities to address guests' recreational and tourism needs. These may include public parks, forests, natural areas, wildlife management areas, lakes and reservoirs, historic resources, trails, rivers, beaches, water access areas, among others.
Research	A detailed and careful consideration of particular concern or problem to find out more about it or a systematic inquiry to describe, explain, predict, and control the observed phenomenon . It involves inductive and deductive methods.
Resource research	Assessing resources available at a destination for visitors. These resources include attractions and services available to support a visitor's stay at a place.
Routing	The process of creating the most cost-effective route through the minimisation of distance or travel time necessary to reach a set of stopovers in a tour. Routing prescribes the path and sequence of stopovers.
Scheduling	The process of assigning arrival and stayover time at each stop – when will the group arrive at a place, and how long will it stay there.
Thematic interpretation	A communication approach to site interpretation where an interpreter relies on a central theme or message relevant to the visitor for presenting the place.
Visitor research	The study of consumer behaviour of visitors to the monuments, parks, museums, and similar places by the overall market, geographic area or individual visitors.

■ Discussion questions

1 Critically analyse Tilden's principles of interpretation and Sam Ham's framework for thematic interpretation. Do you see any convergence?

2 Identify some local natural and cultural resources. Can any be converted into a tourism product? Develop an interpretive theme for the product.

3 Visit a monument near you. First, identify the main visitor segments and why they visit this destination. Then, identify different themes targeted at these segments to interpret the place.

4 Visit a natural area/ park/ historic site and develop an interpretive plan.

■ Activity

1 Suppose a group of school children is visiting your city/ village for a day. If you have been asked to lead them, design a one-day itinerary for them. Then, complete the routing, scheduling and loading. Next, prepare a briefing as you receive the group at the nearest bus station. Identify a theme for this day trip and develop narrations for different venues. Finally, develop a day-end debriefing.

2 Prepare a note on how ready are you to become a tour guide. What would you do to become a tour guide in your area? Identify the steps.

3 Form a group of three and collect information about a chosen tourist place. Note the sources of information and differentiate folklore from the facts.

References

Ap, J., & Wong, K. K. (2001). Case study on tour guiding: professionalism, issues and problems. *Tourism Management, 22*, 551-563.

Crompton, J. L. (1979). Motivations for pleasure vacation. *Annals of Tourism Research, 6*(4), 408-424. Retrieved 6/29/2020, from https://sciencedirect.com/science/article/abs/pii/0160738379900045

Dann, G. M. (1977). Anomie, ego-enhancement and tourism. *Annals of Tourism Research*(4), 184-194.

de Kadt, E. (1979). *Tourism: Passport to Development.* New York: Oxford University Press.

Farnham-Diggory, S. (1972). *Cognitive Processes in Education.* New York: Harper & Row

Gray, P. (2008, August 20). A brief history of education. *Psychology Today*: https://www.psychologytoday.com/us/blog/freedom-learn/200808/brief-history-education

Ham, S. (2013). *Interpretation: Making a Difference on Purpose.* Golden, Colorado: Fulcrum Publishing.

Jezewski, M. A., & Sotnik, P. (2001). *The rehabilitation service provider as culture broker: Providing culturally competent services to foreign born persons.* Buffalo, New York: Center for International Rehabilitation Research Information and Exchange.

McIntosh, R. W., Goeldner, C. R., & Ritchie, J. R. (1995). *Tourism: Principles, Practices, Philosophies.* New York: John Wiley and Sons.

McKean, P. F. (1976). An anthropological analysis of the culture brokers of Bali: Guides, tourists and Balinese. *Joint UNESCO/IBDR Seminar on the Social and Cultural Impact of Tourism.* Washington: UNESCO.

Nettekoven, L. (1979). Mechanisms of intercultural interaction. In E. J. de Kadt, *Tourism: Passport to Development.* New York: Oxford University Press.

Pearce, P. L. (1982). *The Social Psychology of Tourist Behaviour.* New York: Pergamon.

Tilden, F. (1957). *Interpreting Our Heritage.* North Carolina: University of North Carolina Press.

Urry, J. (1990). *The Tourist Gaze.* London: Sage.

van den Berghe, P. L. (1980). Tourism as ethnic relations: A case study of Cuzco, Peru. *Ethnic and Racial Studies, 3*(4), 375-392.

Veverka, J. (1994). *Interpretive Master Planning.* Helena, MT: Falcon Press.

6 Basics of Tour Guiding

Learning objectives

After studying this chapter, you should be able to understand:

- The general dos and don'ts of tour guiding
- The concept of heritage walks
- How to guide visitors on a walking tour
- The nuances of guiding on a coach
- Some essential tools required by tour guides

So far in the book, we have discussed the theoretical underpinnings for interpretation planning, interpretation, and interpretive narratives. However, on the ground, a more clinical approach is required. There are several situations where guides must lead the group. With this chapter, we begin a discussion on how to guide visitors in basic situations.

One of the most basic formats is to accompany guests as they look around a monument or a natural site. Recently, there have been innovations, and themed walks have become popular among visitors. Transfer of guests from one site to another is routine. During the transfers, professional tour guides have developed techniques to narrate on moving vehicles. This chapter also discusses guiding on a moving vehicle. Conventionally, tour guides lead the group in archaeological sites, historical places, religious sites and museums. Then they may have to accompany guests to a natural site. We discuss all these scenarios in the following chapters.

However, first, we have the general dos and don'ts for tour guides.

Dos and Don'ts

Dos

Tour guides must note down all commitments in the scheduler. Alarms must be set so that no commitment is missed, or one is never even late – it is a cardinal sin. The following are essentials that all guides in all situations must ensure.

A tour guide's commentary should neither be too short nor too long. As discussed earlier, the whole commentary must be woven around a planned interpretive theme. Additionally, a good tour guide will relate commentary with the situation: the audience, the events, etc. Guides must plan for Top Visual Priorities (TVPs). Identify which TVPs are to be brought to the audience's attention and when. TVPs deserve the first mention. Then, of course, some TVPs will spring up spontaneously. Tourists visit a destination to immerse themselves in the local experience. So, talk about the social saga and connect with contemporary community life. Tell about life, how people have lived here over the years: their values, beliefs and traditions.

There should be a provision to accommodate last-minute changes. Occasionally, guests may have some special requests. Tour guides must adhere to the daily visit list and achieve the itinerary, however, adaptability is crucial. Therefore, the rule is first to follow the list of things already committed to the guests and then accommodate their requests wherever possible.

Tour guides may be required to change the earlier planned routes or itinerary. For example, guests may like to spend more time some place. Advise them that this would mean cutting time allotted to other attractions. Alternatively, the guide may spend more time with the guests than planned. A good tour guide will always suggest alternative routes and itineraries; however, this should not be an additional cost for the company. Always seek prior approval from the operations team and keep the ground crew in the loop. Similarly, the tour guide may have to alter their narrative according to the audience: adults, youth, children, professionals, etc.

6

Follow the Rule of Three. Tell about not more than three things at a time. Do not get into the details of everything and unnecessarily stretch the commentary. Instead, say what is important and likely to excite the visitors' inquisitiveness. Remember, only the 'should tell' and 'must tell' begin. Then, if the guests appear interested, elaborate with 'could tell'. Avoid technical terms and jargon, especially if they are in vernacular. If a technical term is used, speak out slowly and syllable-by-syllable.

A tour guide must not be restless or in haste. Instead, they must allow guests to talk and interpret for themselves. Offer them the opportunity to put across their understanding of things and events. This exchange, in turn, allows tour guides the opportunity to improvise it with more information and interpretation. Encourage and handle questions (we will discuss this later in detail). Guides must always appreciate it when guests ask questions. A tour guide is not supposed to know everything and answer every question. If you are unsure about the answer, tell the group that you do not know and but you will find out.

Handling a group where the guests are not known to each other requires concern on a few more things. First, always make each group member feel comfortable and part of the commentary. They might be unknown to each other, but you must make them feel a part of the group with your efforts. In a scenario where the guests are from different backgrounds, they may have varying tastes and wander. Therefore, make sure that the entire group is together. Allow them time later to do things they desire and miss out on. Regularly count the members of the group. Then, repeatedly ensure that every member of the group is there.

A tour guide should be confident of their strengths: Voice, persona, humour, knowledge and demeanour! Appear confident and well informed. As a professional, learn to hide nervousness. The guide is the local person and knows the most about the place and subject. Tour guides must continuously innovate. They must design and offer unique experiences for the guests. Thus, they may strive to bring in elements that actively engage the guests. Anticipate what the visitors would like to see and know. Find out if there are any particular likes and dislikes and try to accommodate them. Keep the excitement alive. Present the place or experiences with much energy and in the most exciting way.

There will always be some unexpected situations. A tour guide must be spontaneous and forthcoming to handle these. Later in the book, we discuss some difficult situations the tour guides will encounter. One needs to be tactful in an awkward situation.

Don'ts

A guide must avoid talking indiscriminately. A guide's job is no longer just about information dissemination. It is more about provocation and facilitation. Tour guides should avoid extreme assumptions – either that the visitors already know, so they should not be told; or they do not know anything, and everything must be told in detail. Sometimes guides believe that they must deliver the prepared narration (even when the audience is not interested), or sometimes they tend to compete with some guests by evidencing their expertise (to regain control over the situation). This is not desirable. If someone is particularly interested, the tour guide may explain it to him separately some other time.

Guides should not get self-justifying and use phrases that build up the unnecessary defence: 'I was asked to accompany you on the last moment...', or 'I am not an expert...but will try my best...', or 'If we had more time, I could have explained you better...', etc. Avoid negativity and do not be apologetic: 'Our village is dirty, not like yours...', 'Avoid these beggars, they are not good...', 'I am sorry, the roads are bad...'

They should avoid cynicism and maintain neutrality at all times. For example, in the case of an opinion-based issue, declare that it is only an opinion (and that it hardly matters). They should prefer not to discuss sensitive political philosophy, religious priorities, sexual orientations, etc. Guides should especially take care of physical proximity. As discussed in Chapter 3, the use of space could be misconstrued. Avoid sexual advances, even a discussion on sex.

Never be impolite or discourteous with guests. Never compare them unfavourably to previous groups – never upset them. Likewise, a tour guide should not misbehave with vendors, drivers, traffic, and even beggars or vagabonds. This behaviour creates a poor impression and spoils the mood. So do not create unpleasant scenes.

As for the technique, a guide must not talk while walking or even before everyone has gathered and is ready to listen to them. Guides must never cook up stories and present folklore or fiction as facts. They must not tell what is not authenticated. Instead, they must declare that it is folklore or a popular belief.

The guide must be conscious that they do not prioritise any guests ignoring the others. It creates a sense of inequity among the group.

Finally, never give medication to the guests, but seek expert advice. Guides should never ask for a tip.

Heritage walks

Heritage is an abstract concept. A place's heritage is a sum of all that a community inherits from the past and often transcends into the present. It includes both animate and inanimate elements that reflect the essence of a place as a living entity. The idea and identity of the place include both natural and built heritage and their organic connections to the community. Experiencing the place in the backdrop of its heritage is an exciting proposition for the visitors/tourists (Gentry, 2007). Many of whom would like to internalise the essence of this place. A heritage walk, or merely a 'walk', is an effective way.

Heritage walks bring alive all the diverse elements of the place's living heritage. This is manifested in various elements that include traditional places of worship, monuments, contemporary buildings, lifestyle, traditions, and other cultural imprints. A leisurely walk allows the guest to engage with the community and its properties. A heritage walk is, therefore, meandering through the reality of the place and feeling it for oneself. This experience is often not a single simple thing. The reality of a place is multidimensional, and therefore heritage walks are often designed around a theme (Saiyed et al., 2016) which ranges from a simple tour of the place to special themed tours like a historical event, the heritage of a segment (e.g. Jewish settlement or events of 1857 revolt or the cuisine, period architecture, or something similar).

Heritage walks are becoming increasingly popular for their efficacy of quality of experience and as a tool to impart knowledge about the heritage and history of a place (Chauhan & Anand, 2021). Creative ideas and themes are keys to exciting and enriching experiences for visitors. However, a heritage walk has to be designed professionally with clear objectives and targeted at specific audiences. They should be designed for visitor convenience and use multiple props to enrich the experience. Like any good tourism product, a heritage walk should begin with proper research followed by creativity, planning, execution and room for product improvement.

For example, India City Walks, a walking tour company in Delhi, has developed *Metro Walks* that showcase Delhi's built heritage and cultural traditions. Another popular walk is *Old Delhi Bazaar Photo Walk,* which is also reported as an approach towards low-carbon tourism (Chandran et al., 2019). Similarly, the *Salaam Baalak Trust City Walk* is conducted by kids who lived on the railway platforms before being rescued and rehabilitated by the Salaam Balak NGO. In most cases, these walks have an underlying theme for the visitors. *Alternative London* is a walking tour through London's creative East End. A professional leader helps visitors use cameras during the *Vancouver Photowalks*. *Melbourne by Foot* is about the city's history, street art and cafe culture. Guided tours of the world's largest slum in Mumbai by Reality Tours and Travels are top-rated.

The idea stages

The heritage walk should be developed around a good idea that would set it apart from other competing experience products. Given the backdrop of a place, developers must identify a theme that showcases some aspect of the community's life that should interest the visitors. The walk should allow a peep into an aspect of life that is otherwise overlooked but is an essential component of the personality of the place. Talk to the locals. Search for an idea! For instance, the Chandni Chowk tour is an experience that is unique for foreigners. This tour is partly on foot and partly on rickshaws to provide an essence of the life of locals. Similarly, a *Ghost Tour* of Zurich allows a different experience for the tourist where they can listen to horror stories and visit alleged haunted sites of Zurich.

Some variants of walks include:

- ***Conventional walks*** are walks through the built heritage, passing through the community's monuments and other built assets. They revolve around the history of the place.

- ***Non-conventional walks*** show a non-conventional aspect of the community. Examples abound in slum tours, street food walks, spice market walks, textile walks, or more recently, walks following the election campaigns of political parties.
- ***Theme walks*** are woven around a theme ranging from a food trail to a historical event to Galib trail (of Delhi Metro walks). Another example is the *Game of Thrones Tour* in Dubrovnik allows tourists to see the sites where the popular series was shot, giving time for a good photo session.
- ***Period walks*** are walking through the story of a period in history, for example, the Mughal period or the Colonial period. These walks usually showcase period architecture.
- ***Contemporary walks*** showcase the life and time of a present-day community, for example a walk through the traditional settlement in the old town, or walking through the community as it readies for some event/ festivity or just a shopping trail in a traditional bazaar.
- ***Spiritual walks*** are walks through the places of worship that the community identifies with, in past or present.

What is a walk?

The walk is essentially a stroll through and around a set of cultural/ heritage elements typically along a predetermined route with identified stops and opportunities for interpretation, maybe along an underlying theme. Following are the steps to designing a heritage walk. These steps have to be dealt with iteratively. You may start with Step 1, go to Step 2 and come back to one

6

Step 1: WHAT?

Having decided on an idea for a walk, one must decide what the walk would like to achieve, and at whom the walk is targeted. This focus will also determine which elements will be chosen for interpretation. Some examples of objectives of the walk could include:

- What are the daily habits of the locals?
- Why the local architecture is the way it is? Why and how a particular building material will be used.
- How was it during the 1857 revolt?
- How did a particular community live in a particular historical period?
- What is available, and how do the community connect to the street food?
- How do people shop around in routine?

Once the 'what is to be done' is decided, the next step is to identify the resources for interpretations.

Step 2: HOW?

- Begin with a map of the place.
- Identify the resources that fit the walk's theme; locate them on the map
- Connect them with a route – consider its length, the flow of the story, time available for the walk, etc.
- Identify the starting point and the terminal point. The starting point should be easily accessible and readily found for the tourists.
- Identify stops and activities along the route, and the artefacts that best make a point (what you would like to show).
- Identify the concerns – safety, rugged terrain, weather, narrow alleys, traffic, etc.
- Evaluate how the community that would be trespassed would connect/ react to the walk. How can the perceived intrusion be minimised?
- Evaluate if the routes can be standardised and stops can be marked.

Some of the design considerations for a heritage walk are:

- A walk should be around an hour or so long.
- It should not be more than one and a half km on foot; maybe a bit longer if some alternate travel modes are used (note the use of Segways for city trails or use of rickshaws, etc.)
- Plan for 8 to 10 stops on the way. Each stop may be a landmark in the story. Assess the convenience of stopping at the chosen point (congestion, space, traffic, etc.).
- Plan for activities and add-ons. These add-ons may range from shopping to tasting food or participating in a ritual of activity.
- Consider possibilities of interaction with the host community (participating is a community activity or meeting a family of an elder/ leader).
- The walk should provide tourists with an experience which they cannot have on their own. Otherwise, the efforts will be futile on both ends.
- The visitors must be aware of the amount of walking involved, flights of stairs, and any attire they must wear, particularly for a visit (to shrines, temples etc.)
- The guide must inform them about the weather and environment in advance.

Guiding on a walking tour

A walking tour is the most elementary form of an organised tour. It assumes a different character depending upon how a walk is envisaged. For example, there could be one to two-hour stand-alone tours at a destination. Some of these are included in the package bought by the group. There will be an expert leader for the walk whom you should contact and engage for leading the walk. Alternatively, a short walking tour at an attraction can be included as a part of the more extensive itinerary where the role of the tour guide is to take a party around as and when it arrives at the site. Nowadays, walking tours are becoming popular as heritage walks or city walks. Tour companies, these days, are offering added value for guests through innovative themed walks. Whatever be the nature of the walk, basic principles of leading a walk should be attended to.

Preparing for the walk

As discussed in Chapter 5, the key to any successful walk is thorough research. Tour guides need to research history, folklore, anecdotes, special interests, etc. and familiarise themselves with local names, traditions, terminology, etc, as well as the attractions, site transport, public authorities, hospitals, hotels, restaurants, local performers, etc.

Immediately before the walk

The first thing is to identify an appropriate point to initiate the walk. Consider the size of the group and other groups present in the vicinity. If the group is arriving by a coach, such a point should be preferably near the parking or at a stop nearby. Give the group contact person clear instructions about the meeting place. As the group arrives, gather everyone and count. Confirm if everybody is there and cognised. Always let them know that if for any reason they have to leave the group in between, they must meet at a place and time set by you.

Tour guides should be calm, composed, and confident as they take charge of the group. The next thing is to assess the group's abilities to negotiate the proposed route and itinerary. Check if there are any members with special needs. Finally, tour guides must wait for the entire group to be ready before starting.

As you begin the walk

Introduce yourself and brief the group. Describe the tour succinctly. First, share the route, perhaps with a map, showing the distance traversed, stops and end point. Hand out copies of the map if useful. Describe the level of difficulty that will have to be negotiated and warn about anticipated challenges, including steep climbs, uneven surface, crossings, adverse climate, and others, if any. It is essential to advise the members to locate the party or get back if separated from the group. Similarly, if a person can not manage a section of the route, e.g. a steep flight of stairs, they need to be guided on where to wait for the group. If the group is touring the city or village, it needs to be sensitised about the host community. These things are essential to be told in the beginning.

At this point, the tour guide must also brief the guests about safety issues – appropriate dress and footwear; traffic rules, especially about the orientation (walking on the right side or the left side) as people from different countries are used to different norms; regulations like no smoking in public places; and concerns while walking through the congested areas like remaining aware of unsocial elements, if necessary.

The tour manager will invariably be the 'sweep' if a local guide leads the group. The job of a sweep is to be the last member of the group, ensuring that all party members are always with them. The tour guide may designate one guest as 'sweep' if the tour manager is unavailable. However, the tour guide should not rely too much on this sweep, as they are also there to leisurely enjoy. It is the guide's responsibility to ensure that all members are there.

On the tour

As the group starts the walk, the guide must always allow them to catch up, especially as they enter a monument, building, structure ora site. Welcome all questions and remain observant and point out and explain TVPs. Ensure all group members have gathered around you to listen to you. The guide must never walk backwards facing the group (danger of stumbling and falling).

If a question is asked, stop and call everyone's attention to the issue. Repeat the question loudly so that everyone can hear it clearly and then answer it. Watch for uneven surfaces, low-height doors, and other challenges. Keep informing the group: 'please watch your steps...' or 'please mind your head...', etc.

Considerations in walk tour commentary

Three crucial considerations in the walking tour are 'pace', talking to group' and 'positioning'.

Pace

The guide must carefully set the pace of the outing according to the group's capabilities. They must estimate the physical capacity of the group members. Adjust the pace of the walk according to the slowest member of the group, ensuring that no one is left behind. They must keep an eye on the sweep, and must ensure that all members are between the sweep and themself.

While walking in the streets, guides must be vigilant as they take a turn on a street corner. They may wait until everyone has closed in and are ready to follow them on the turn. They must not stop abruptly, especially when the group members closely follow them. Similarly, do not stop very frequently. Frequent 'walks-stops-walks' disrupt the momentum of the walk. If there are many stops in quick succession, then walk slowly.

Talking to the group

As discussed in Chapter 4, one of the basic narration rules is not to talk as you walk. Group members follow the guide as they walk, and not all members are close enough to hear the tour guide. Otherwise, you will repeat yourself constantly, creating irritation to the group and yourself. In addition, when they see the guide speak, they may rush in and failing to do so may disappoint them.

Every time the group assembles/reassembles, glance through to ensure everyone has made it to the group, establish eye contact, and start speaking. Always convey through your eyes that you are ensuring that everyone is there. As a result, members will feel happy and satisfied. They will believe that you are doing a fair job and ensuring that they do not miss anything – and so there is no need to rush and run.

Positioning

The third consideration is positioning. Positioning refers to the group's formation while listening to the tour guide. Guides may choose a stop at a location where your group can assemble conveniently. Ensure that there is a large enough place to accommodate all. When in the open, look for a sheltered/covered place, especially if it is too much sun, rain or wind).

As advised in Chapter 4, the group should assemble in a small semi-circular formation, such that the guide faces them, but to the side of the object or site to be viewed. The guide may gently (politely) point as they talk to indicate to the guest who inadvertently moves out of their sight to get back in.

It must be ensured that the group does not block an entrance or walkway, and that guests do not stand in front of the object or the exhibit for an extended commentary as other guests or groups might have been waiting for their turn. If another group is around with their guide, try to organise the group so that the speakers are back-to-back, and their group faces them, which means the other group is almost 180° away/apart. This will help prevent the mixing of the two speakers' voice-in-commentary.

Other considerations include ensuring that group members do not have to look into the sun. Similarly, the guide should not stand and speak against the strong wind/breeze. Instead, they should use the flow of the breeze to help carry their voice to the listeners – position the group downwind.

Do not ask the group to turn abruptly to back and then to the left and right again, etc. Frequent turns are perceived as abrupt and break the momentum of the group. Organise your commentary so that the group view traverses from one point gradually moving in a clockwise or anti-clockwise direction without having to turn back frequently and suddenly.

Gigantic structures like the elevation of a fort gate or a minaret are best viewed from a distance. If the structure is in a lane or a street, it will be best seen from across the street. After that, move close to show the finer architectural details. Use semi-circular formations in a triangular position for this. Guides must point out objects precisely: 'the three-story Georgian building on your right is where the Governor of the county lived...'

Make sure that the group does not assemble in the middle of a path or courtyard blocking others. Instead, encourage them to stand on a side, say, close to a wall or a barricade that serves as a trammel. If there is a narrow lane which will make it difficult for the group to assemble, stop short of it and tell the group ahead that 'we will pass by a dilapidated house on your left with a big green door and huge window – that was where the famous courtesan lived'. Lead the group to pass by the structure slowly without stopping and causing congestion.

End of the walk

As the walk comes to an end, ensure that everyone is there. Debrief the group, which might include summarising the tour for a minute or so. Use this opportunity to orient the group about the next leg of the tour and provide necessary information and instructions. If there was local assistance like a museum docent or a site caretaker, they might be thanked at this stage. If the trip ends, this is the time to exchange pleasantries, say goodbye, accept gratuities and descend into some informality indicating that the tour is finally over.

Important precautions

Walks are typically organised in public spaces where there will generally be other visiting groups, locals, stakeholders and their clients, and authority executives. A responsible tour is when visitors' movements do not obstruct others.

When a group is in a public place, the tour guide must remain watchful of the safety and security of the guests. Sometimes tagalongs may get associated with the group. Guides may politely ask them to disengage.

There will be distractions as well. For example, it could be some of the guests themselves. Professional tour guides handle these guests, not allowing them to disrupt or distract the interpretation. Even the photo-snapping members of the group should not disturb the flow of the commentary or tour.

Delhi Walk essentials

- Walks in Delhi are partly on foot, on a rickshaw and by metro rail.
- Guests should have good walking shoes that can be taken off and carried at religious sites.
- The guide should instruct guests to carry a headcover to avoid the sun. It may be necessary to cover the head at some sites of worship.
- The guide should ask guests to wear a mask in crowded places. They should also cover their mouth during a visit to the spice market
- Guests should carry enough drinking water.
- Dress unobtrusively. No shorts/ half pants will be permitted in religious sites. No leather items are allowed in the Jain temples.
- The average time per walk is usually around 2-3 hours
- Some walks are scheduled in the early mornings or evenings to suit the climate and place.
- Feel free to leave a donation at religious sites.
- Carry small changes for the metro and rickshaw rides.

Leading on a moving vehicle

One of the most common situations is handling a group on a moving vehicle, generally a coach. Usually, the coaches are pre-organised by the tour operator. The driver is responsible for the bus and drives the group through /to the destination. The tour guide, representing the tour company, is in charge of the tour and must work with the driver as a team, with the guide soliciting the driver's cooperation with routes, stops, parking, and other arrangements. Guides need to empathise with drivers as they deal with traffic, regulations and manoeuvring a large vehicle. Drivers are often called 'pilots' or 'coach captains'.

Most of the techniques of a good narration are applicable. However, there are certain things to note.

■ Preparing for the coach ride

The tour guide should connect with the company providing the coach. First, they must talk to the tour manager in the travel agency to learn which company is providing the coach; also, confirm the size of the group to determine the size of the coach required. Then they must talk to the coach provider to have the details of the coach – coach number, size and make, the name and mobile number of the captain.

They must take note of the route to be followed. Plan the drop-off and pick-up locations at the site, considering the parking space at these points. Check if the coach cannot stop or travel to some points. Also check, if any of the routes are blocked for the day, and work out alternative routes if necessary. Prepare notes: identify the main buildings and structures to point out on the route. Learn about their history and current use. Often the transit time in a coach is when guests want to know more about the place. So, a tour guide must collect exciting facts about the city, its life, history, and demographics, among other things, to be used as fillers.

Immediately before the ride

The tour guide must arrive early, at least 15 to 20 minutes, to locate the coach and meet the captain. They must introduce themselves to the captain if meeting for the first time and exchange pleasantries. If this is the first meeting, note his/her name and name to be addressed and his/her alternative cell phone number, if any. Discuss itinerary, route, time and timing of stops and other details with the captain. Ask for the rules inside the bus and brief the guests accordingly. For instance, ask if guests are allowed to eat and drink inside the coach.

Before the ride, the tour guides must ensure that the coach is in good condition. First, they must thoroughly inspect the bus physically. Check with the captain for broken panes (if any), seats, emergency exit, hand brake, fuel and other details. Next, they should locate the microphone and test it. Check if the audio system is working correctly. Next, disinfect the mouthpiece with a sanitiser. Finally, ensure that the first aid kit, fire extinguisher, and other safety gears are in place; locate emergency doors and windows.

Claim the seat with access to the microphone. The guides may place their bag or cap, or other possessions on the seat to indicate prerogative.

About to start the ride

As the guests arrive to board the coach, the guide must receive them at the coach door with a welcoming smile and count the heads (match them with the naming list you have) to ensure everyone has boarded and is comfortably seated. Then they must ensure that everything is firmly in place. Finally, the guide must ensure that the objects are safely stacked in overheads or luggage storage racks/bins. Also, check for audibility, asking guests if they can clearly hear him/ her – the volume and tone – otherwise, make necessary adjustments.

Begin by introducing the team – the captain and the assistants (if any). Do not refer to him as 'driver'. Call the driver a 'captain' or a 'pilot'. Introduce any other staff/help. His assistant may be called 'Assistant'. A word of admiration for the driver and his assistant would go a long way in keeping up their morale.

Advise guests to note down the coach number and the name of the company (painted on the coach body). Then, ask them if they carry things like water, necessary medicines, an umbrella, and other essentials. Check if the group has the necessary tickets and permissions. Finally, politely ask them to not take the mobile number of the driver as the tour guide will be the primary contact.

Brief the guests about the programme/itinerary. Instruct where and when the tour will end. Remind the group about the need to adhere to safety procedures. For example, passengers inside the coach must fasten belts in some countries. Instruct the guests that there should be no walking in the aisle whilst the bus moves. It is for their safety and wellbeing. Point out the location of emergency exits, emergency glass breaker hammer and seat belt cutters, fire extinguisher, and other safety devices, and show guests where the first-aid kit is kept.

Usually, there is no enforceable policy in many countries about eating and drinking on board. However, request guests to keep the coach clean and avoid eating and drinking. Finally, explain and announce to the group that the driver and the guide are there to help and may the trip be as pleasurable as possible for them. Should they require any help, they may let the guide know.

On the tour

Typically, coaches have two different orientations. There is a special seat for the tour guide that faces the tourists in some, while in others, the first seat near the door is earmarked for the tour guide. If the guide sits forward-facing with their back to the guests, then the guide's right and left are the same as guests', making it easier for them to give direction. Occasionally, the guide may turn back to glance at the guests to establish eye contact and assess their interests and dislikes.

However, if the guide is sitting facing the guests, then the guide's right is guests' left, and the guide's left is their right. Therefore, the guide may note and mentally prepare according to the seat arrangement. To avoid confusion, always raise your said hand while giving directions.

While pointing out objects and TVPs, always say 'to your right...' or 'to your left...' Never say 'to my left...' or 'to my right...' It has to be the guests' right or left. Similarly, do not give directions vaguely, such as 'here', 'over there', 'on the corner' or 'at the back'. Instead, use descriptions to identify what is to be seen and where the visitor should look. e.g. 'the red sandstone building with massive cast iron grill gates to your right is the palace of the royal family.' Also, be aware of not using more than two languages simultaneously unless essential.

Professional guides must think ahead. They know your route and therefore are aware of what is approaching. Point out objects of interest and TVPs just as the coach approaches them, allowing guests to get ready to see what is being said. Generally, point out what all guests on the coach can see. Do not point out to an object or TVP as the coach passes by or has passed by. It is useless to say, 'we've just passed by the statue of the king who founded this city...' If the group gets down from the coach for a while, do not forget to tell them where the bus will be parked and how they can locate it. Give necessary instructions to reassemble on time and at the agreed meeting point.

Remain watchful of ambient temperature, and ask the captain to adjust the it to guests' comfort. Also, keep a watch on the driver and do not allow them to drive rashly. It has to be a relatively slow and smooth ride within cities.

Play music only when asked. Ask guests about their choice and play accordingly. Sometimes a guide would like to play local music to educate the guests.

Using the microphone on the moving vehicle

Tour guides must master using a microphone on a moving vehicle. Microphones these days are sensitive but need to be supported by a good audio system. While using a microphone, the guide must speak unhurriedly and clearly, with normal voice, tone, volume. You need not make an extra effort because it is not required, and you will not be able to keep up it throughout the trip.

Hold the microphone close to your mouth. Keep it on the chin just below the lips for support, and so that the distance between the microphone and the mouth does not frequently change so that the voice is not lost intermittently. If you are irritated by the feel of the microphone, place your thumb firmly between the microphone and your chin. It might be unhygienic to put the microphone used by others so close to your mouth, and so it is advised to either cleanse it with sanitiser before use or carry a personal sponge swathe (cover) for the microphone. Tour guides should avoid sitting or standing below an amplifier as this may cause noise because of electromagnetic wave interference.

End of the tour

Often the tour come to an end with the coach arriving at the hotel or other designated point. It is an opportunity for debriefing. Summarise the activities. Share the plan for the next day or the next part of the tour. Also, provide information on any relevant group activities following your tour. For example, 'you're all getting down at the market. After shopping, you can take a metro from here for route no 55, which takes you to your hotel....'

Say a general goodbye to the group over the microphone or publicly. Visibly thank the captain and other staff. See if any tips are to be given to them. For guests, remind them not to leave anything behind. Request them to check overhead storage and floors for personal belongings. Demonstrate concern by advising guests to mind their heads or steps as they disembark. Stand close to the gate and exchange pleasantries individually as members of the group get down one by one.

Once the group has left, help the driver check the coach and ensure no guest has left anything behind. Finally, debrief with the driver.

■ Important precautions

While professional tour guides must perform and keep the interpretation enjoyable, still they need not be rock stars. Do not carry the cabled microphone all over the coach and speak. A guide must not stand or walk and talk on a moving vehicle. Guest may keep watching the guides distracting posture or balancing act instead of listening to them. They will become the TVP. Guides are not TV presenters.Do not copy them; instead, be yourself. Like singers, do not hold the microphone in front of the mouth – it hides the face and restricts eye contact.

While transit on the coach is an excellent time to share information about the place and forthcoming experiences, guides should be mindful of the situation and engage in calculated talking. Do not chatter all the time. Do not talk with the captain while the vehicle is moving. If the captain is to be directed, it should be a short to the point instruction. Avoid constantly asking the driver how far the destination is. If you need an update, use GPS technology.

Guides must also be careful not to chew candies, gums, or tobacco as they talk. Do not always keep the microphone 'on'. Put it on only when there is something to say. Do not make loose comments on the group, a guest or otherwise – the microphone might have been inadvertently 'on'.

Some guests rush to capture a prime seat on the coach – other guests who are not agile feel deprived. As discussed earlier, this may be perceived by them as inequity. The tour guide must ensure that no guest patronises any seat. Set a roster and request them to rotate so that everyone gets a fair chance to be in a place considered privileged or otherwise.

Tools of the trade

Every professional tour guide would require a few essential things in his/ her kit. Likewise, a few things would come in handy for a tour guide.

Certification

Only an authorised individual can lead the group in many countries and sites. Guides must check if they require a certificate or a license to operate. In some cases, a tour guide is mandated to carry the badge or license on the body. Any person leading tourists may be asked to show their license to the authorities. Guides must oblige.

Smartphone

Technology plays an essential role in contemporary tour guiding (Pirolo & Nasta, 2020). Therefore, guides must leverage technology to improve their efficiency and effectiveness. In the past, tour guides used to carry a lot of gadgets and tools for support. However, a smartphone is almost the entire office on the move these days. It keeps a tour guide connected with concerned people – the road crew, the tour operator, the monument authorities, the driver, the restaurateurs, and other stakeholders. Smartphones also double up as an address book for the local contacts.

Smartphones today are loaded with helpful apps. They are the route planner, information storage, camera, torch, and other utilities. If it is a multi-day assignment, the tour guide may create an e-group for swift communication with guests.

A tour leader must always carry necessary maps with them. They may need to show these maps to the guests to give them an idea of the destination and location, mainly when the attraction is spread over an extended area. Smartphones have GPS driven apps like Google Maps or Polaris GPS navigation to assist tour guides and guests.

You may be a linguist. However, there may be a situation when a tourist needs some explanation, and you would like to provide a translation that is not coming to your mind readily. There are many foreign language translation apps. Similarly, inbuilt scanners can read a signboard, a description or instruction in the local language and translate them. In earlier times, guides would carry foreign language phrasebooks.

Similarly, earlier, guides used to carry a torch with them as they may have to lead tourists to dark basements and places where enough light is not there. They may also be escorting a group in the evening/night to a place where there is no power or a power cut. These days all smartphones have an inbuilt torch.

The phone is also the camera. Guides must also learn some basic photography which is assistance to their guests. Sometimes guests would want the tour guide to snap a few photos and share them back. Some companies want guides to capture some happy moments with guests to use for a tour-end souvenir.

Many apps support video players, animations and presentations. Tour guides will often use multimedia players on their phones to explain and describe what is not explicit. For example, a tour guide may carry old pictures of the site or some exhibits that are not there; or they may also like to share graphic descriptions, layouts, or renders.

Guides often encounter things and situations they may want to note for future reference. Of course, phones also can take notes. Guides may have to note down complaints or unanswered questions that they may like to answer later. They may also like to keep track of complaints.

Many downloadable apps are available to support tour guides. These include safety apps like Accuweather that makes the latest weather reports and current conditions available. Some apps offer assistance in first aid, including CPR. Finally, they have offline survival manuals, which are very useful during outdoor activities.

Many independent tour guides rely on social media like Facebook, Instagram and Twitter to promote their tours. Apps like Trip Journal help guides record, document and share their trips with potential clients. Independent guides require booking and payment applications, and need apps to help them book tours on the go and make changes if required. These apps help guides receive payments, check-in guests on arrival and validate tickets using QR codes. The app may also be accessible to the tour company or partners to check availability and make book tours for the guides.

Many tour companies have created apps to be used by guides while they handle their clients. Guides are required to report the progress of the tour intermittently. In addition, they must update the company on achieving key milestones and events, and on untoward incidents. They are also required to provide necessary feedback on the groups' preferences, behaviour and conduct.

Accessories

Raingear and umbrella

In rainy weather, tour guides are expected to take necessary precautions, and carry rain gear or umbrellas. In addition, they might have to provide cover to guests and may have to make some arrangements still while it is raining.

Similarly, an umbrella is handy when a walk is in the scorching sun. A colourful umbrella may also double up as a prop which members of the group may see from a distance and when in a crowded place.

Cap and sunglasses

Tour guides may have to go out in the sun. As a result, their head and hair are exposed to sunlight for an extended period. This exposure may be incredibly demanding in a hot climate. The tour guide may therefore carry a cap. The cap may be of a fluorescent colour so that it is visible from a distance and is distinguishable in the crowd.

For the same reasons, tour guides may carry sunglasses. However, they must take off the sunglasses as they speak to ensure eye contact. Nevertheless, they may wear sunglasses while moving from one point to another.

Portable voice amplifier

Some tour guides have a feeble voice, or they do not want to put extra effort into raising their volume while modulating their voice. This tool is more desirable for guides who lead several groups in a day or lead at busy locations. Sometimes they are not audible, not understood, or may have to repeat what they say. Therefore, it is increasingly becoming commonplace for guides to use a portable voice amplifier. They use a lavalier microphone or lapel mic, or a headband mic, allowing them to keep their hands free to gesture naturally. It also helps prevent background noises from being picked up by the microphone. Moreover, given the COVID situation, a tour guide can use this amplifier to speak clearly through the mask. The speaker of the address system can be worn on the adjustable neck or waistband strap.

Backup/ charger/ cables

One of the essential tools for tour guides is their smartphone which is used extensively. It must therefore be ensured that the phone is sufficiently charged. In addition, most vehicles have USB ports for charging phones and other gadgets. Therefore, guides must essentially carry chargers and cables with them. It is advisable that they also carry with them a power backup.

Props

Tour guides might carry a placard or a small brightly coloured flag to indicate your location in a crowded place. This prop will help guests who might feel lost.

First-aid kit

Guides are not advised to double up as doctors. However, as a precautionary measure, they must carry first aid supplies like cotton, band-aid, spirit, and some essential medicines in their tool kit.

Essentials

Water

Tour guides are expected to do much talking and may require water. Especially in hot weather, water intake increases. A water bottle is essential.

Breath mint

Tour guides often speak to guests from close quarters. Therefore, their breath must not smell foul. If you doubt a foul breath, it is always advisable to keep and use breath mints.

Cash

Guides must always keep some money in cash. While e-transactions are rising worldwide, one never knows when they require cash. They may have to buy tickets, vouchers or pay for unforeseen expenses.

Necessary documents

Guides may be required to carry documents on behalf of the group to ensure smooth travel and transit. These may include itineraries, vouchers or money for entrances, confirmations for hotels/restaurants, rooming lists, entrance information, among other things. These days, guides may access all this from their smartphones.

Towels, wet tissues, and deodorant

In hot and humid parts of the world guides might sweat on a long day or a challenging track. Thus, they may keep a small towel or wet tissues, and as and when required, they may wipe their face.

If the tour guide sweats a lot and smells, the guests may not like to come closer. Nothing can be more embarrassing for the guides. Therefore, they must use a deodorant to prevent chafing.

Sanitiser, disinfectant, and masks

Post-COVID guides must carry sanitiser, soap, and disinfectants. Guides must frequently wash their hands. They need to touch many surfaces and disinfect them before using/ touching them. They must also wear masks at all times.

Things to avoid

Finally, let us look at a few things that one may detest as a professional tour guide. It is also essential to keep the backpack as light as possible.

Notes

An amateur tour guide may prepare some cards to note a few key points about the destination. They are tempted to use them as cue cards. However, it must be borne in mind that guides should never take out their cards and read from them. Their narration has to be spontaneous. Never have a canned speech or one that sounds like it is coming out of a textbook.

Personal audio

The guests are on holiday, but the guides are not – they are on the job. Guides must never convey a casual attitude. There might be occasions (especially on the coach) when they are not speaking for some time. It is advised that guides do not plug in their earphones. Tour guides should avoid taking personal music/ video with them.

Cigarettes, tobacco, and alcohol

Smoking is injurious to health. In many parts of the world, smoking in public places is illegal. It would be almost a crime to smoke cigarettes when with a group of tourists. Guides should resist the temptation even if they are offered a cigarette by one of the group members. Similarly, guides must also ensure that they are not under the influence of alcohol when on duty.

Key points

In this chapter, we discuss some of the basics of tour guiding. There are certain dos and do nots for tour guides applicable in most situations. Some of the dos include prioritising TVPs, being flexible to accommodate, as far as possible, last minutes demands of the guests, facilitating experiences; and performing like a professional, among other things. Guides should speak selectively, avoid cynicism and any discussion on sensitive issues, and avoid canned narrations.

There is a growing interest in experiences through heritage walks, which bring alive the diverse elements of a place's living heritage. A leisurely walk allows a pace to the guest that facilitates engagement with the community and its properties. Heritage walks must be planned with clear objectives and targeted at specific audiences. They should be designed for visitor convenience and use multiple props to enrich the experience. There are different types of themed heritage walks.

Guiding on a walking tour is one of the basic tour guiding activities. Guides should prepare well before the walk. First, they must carry out adequate research and familiarise themselves with the place. Then, a few preparations are required immediately before setting on the walk.

There are three crucial considerations in the walking tours. They are 'pace', 'talking to group' and 'positioning'. Pace refers to the speed at which the tour takes place. Positioning relates to the group's formation while listening to the tour guide. At the end of the tour, the guide must debrief the group, exchange pleasantries, say goodbye, and accept gratuities.

Another common situation for tour guides is leading a group on a moving vehicle, as visitors travel from one site to another or go around the city. The accompanying tour guides interpret the place for the coach bound guests. This guiding on the coach requires some preparation beforehand and has special requirements, including handling the microphone. In addition, there are particular concerns that the guides must be aware of while on the coach.

The chapter also discusses some tools that the professional tour guide should carry. These include certification and smartphones. Smartphone plays a vital role in contemporary guiding, and can be loaded with several apps that help them improve efficiency and effectiveness. In addition, guides must carry some accessories that include raingear and umbrella, caps and sunglasses, portable voice amplifier, backup and charger, props and first-aid kit.

Then some very essential things must be carried by tour guides. These include water, breath mints, cash, required documents, towels, wet tissues and deodorants, sanitiser, disinfectants, and masks. Professional tour guides must not use personal audio, cigarettes, tobacco and alcohol.

Key terms

Apps	Applications or software that can be downloaded on a mobile phone and have a specific or narrow function. Tour guides need to have some useful apps on their phone to assist them with safety and security, booking and payments, and social sharing, among other things.
Certification	The process of providing an aspirant with an official document attesting to his or her abilities to accompany and interpret for the visitors. Many destinations only allow certified professionals to guide visitors.
Heritage walk	A slow walking tour led by an expert to interpret structures, culture, tradition, philosophy, myths and associated rituals with them at a destination. A heritage wall can be themed around any or more concepts.
Pace	The speed at which the tour takes place. A professional tour guide must carefully set the pace of the outing according to the group's capabilities.
Portable voice amplifier	Also called a PVA, a compact mobile public address system that the tour guides typically wear on their bodies. It has a lapel or headband microphone, while the speaker is held on an neck or waistband strap.
Position triangle	Refers to a suggested triangular formation while guiding. The tour guide, audience, and the object being interpreted must be on the triangle's three vertices to ensure an unobstructed view to the guests as the guide speaks.
Props	Objects that tour guides use to help them conduct a tour. These props can also double up as a guidepost. It helps the visitor groups to locate the leader and follow him/her.
Smartphone	A wireless handheld phone that has computing capabilities.
Sweep	A designated individual in the tour party whose role is to move last in the group and ensure that everyone is with the group. If there is an accompanying tour manager, he/ she often doubles up as sweep.
TVPs	Top visual priority is on top of visitors' attention over other things in a gaze. A tour guide must always look for top visual priorities. The guide should, on priority, talk about things that have captured visitors' attention.

6

Discussion questions

1 What would you do as a tour guide if poor weather interrupted your planned activities?

2 Design a one-hour walking tour of your city/ village. Create a poster to promote the tour.

3 Prepare a short video introducing yourself and welcoming guests for a themed heritage walk.

4 As you meet the group for the first time, you realise that a few members have difficulty walking. What will you do?

5 Prepare a transcript for a route in your city where you would be guiding the guests on a slow-moving coach sitting facing the front. What precautions would you take?

6 The tour has ended for the day, and the party is returning to the hotel. Debrief the group on the coach. Point out important things to be included in this debriefing.

7 List some apps you would like to download on your phone before you set out to lead a group. Then, briefly discuss each of these apps.

8 It is going to be a long day for you. You have three back-to-back tours booked. So pack your bag for the day. Share what is included in your bag?

9 What are some of the things that tour guides must avoid when leading the group?

Exercise

1 Create a bag for yourself as you go for tour guiding in Lahore during two different seasons – August and January.

References

Chandran, A., Kashyap, S., & Suneeth, B. G. (2019). Low Carbon Tourism: an Explorative Study on Walking Tours Fostering Responsible Tourism in Delhi. *Atna Journal of Tourism Studies,* 14(2), 13-31. doi:10.12727/ajts.22.2

Chauhan, E., & Anand, S. (2021). Guided heritage walks as a tool for inclusive heritage education: a case study of New Delhi. *Journal of Cultural Heritage Management and Sustainable Development,* Ahead of Print. doi:10.1108/JCHMSD-08-2020-0120

Gentry, G. (2007). Walking with the dead: The place of ghost walk tourism in Savannah, Georgia. *Southeastern Geographer,* 47(2), 222-238.

Pirolo, L., & Nasta, L. (2020). The changing face of museum tour guides: How to raise audience interest and engagement through digital technology. In M. Massi, M. Vecco, & Y. Lin, *Digital Transformation in the Cultural and Creative Industries: Production, Consumption and Entrepreneurship in the Digital and Sharing Economy.* London: Routledge.

Saiyed, A. A., Basalingappa, A., & Sinha, P. K. (2016). Value Network in Heritage Walks: Case Studies of Ahmedabad City Walks. *Journal of Heritage Management,* 1(2), 191-204. doi:10.1177/2455929616687897

7 Guiding Scenarios

Learning objectives

After studying this chapter, you should be able to understand:

- How to guide visitors in different scenarios
- What is cultural tourism? Its importance, and visitor motivations and expectation
- The technique of guiding at an archaeological site
- Nuances of guiding at a religious site
- How to lead a group in a museum
- Why visitors are attracted to nature
- Nuances of leading guests in nature

Tour guides often need to interpret cultural and natural settings. In this chapter, we discuss three specific cultural settings – guiding at archaeological sites, religious sites, and museums. We also discuss guiding on nature trails.

Cultural tourism

Cultural tourism is a popular reason for tourism, where visitors engage with local culture, specifically the hosts' lifestyle in their geographical setting. The hosts' lifestyles are often influenced by their history, art and architectural legacy, religion, and other elements. While visitors experience the contemporary, they try to make sense of it by visiting monuments, archaeological sites, museums, and places of worship and religious significance. During a limited period of visits to the destinations, visitors often need experts to interpret these elements.

According to UNWTO (2020), cultural tourism is "*A type of tourism activity in which the visitor's essential motivation is to learn, discover, experience and consume the tangible and intangible cultural attractions/products in a tourism destination. These attractions/products relate to a set of distinctive material, intellectual, spiritual and emotional features of a society that encompasses arts and architecture, historical and cultural heritage, culinary heritage, literature, music, creative industries and the living cultures with their lifestyles, value systems, beliefs and traditions.*"

Cultural tourism can often be slow tourism based in rural areas, where the visitors engage in traditions of indigenous cultural communities to have a feel of their values and lifestyles. It is also commoditised urban tourism to historical and cultural centres where they tend to have a snippet of places' history, connected with the present mediated by an expert – a tour company or an individual. With technological innovations, visitors are increasingly empowered to surf the destination independently. However, this does not undermine the need for and importance of assisted cultural encounters on the tours.

Visitors arrive at a destination for an authentic story rooted in history, tradition, and often the contemporary in life within the physical context of the place. The underlying assumption is that the host community has a story to tell. Destination leaderships organise for personal and non-personal interpretation. The guide is one of the essential components of this scheme of interpretation.

Cultural tours are a big business today. Destinations in many emerging economies rely on cultural tourism for their economic development (OCED, 2009). Major tourist attractions are often historical and cultural places. It is believed that cultural tourists spend more at the destination than other tourists. While some visitors arrive with the sole purpose of cultural tourism, others append cultural tours to their travel plans. For example, an individual on a business trip or visiting a friend or relative might take time for a limited cultural tour. Unintentionally, a visitor may experience cultural immersion by interacting with locals, their language, cuisine, traditions and customs during the stay.

Besides the economic returns, cultural tourism also has far-reaching social impacts. Being the tourism product, the host community becomes more aware of and concerned about their cultural assets. While retaining the cultural identity of these assets, tourism entrepreneurs creatively present them to the world, especially the visitors. They may subtly adapt their offers to be more presentable and consumable. It creates a sense of pride with cognisable attempts to regain and restore the cultural identity. Engagement with visitors from different cultural backgrounds and attempts to construe the culture for them also enhance the cross-cultural sensitivity of the hosts.

Types of cultural tourists

As the industry realised the importance of cultural tourism, there were attempts to understand and typify cultural tourists towards the end of millennia. The purpose of discussing the typology of cultural tourism is that the tour guide should try to understand the underlying motivation for the tour and accordingly have an interpretation plan for them. Some guests will need superficial information which is broader but shallower. Some other guests might look for a narrow yet more profound understanding of the cultural products.

Based on the strength of the drive for consuming cultural tourism products, Bywater (1993) offered a three-way typology of cultural tourists:

- First, while holidaying at a place, some visitors are interested in the local culture, but are not at the destination primarily for this purpose. These are classified as *culturally interested* visitors.
- Some visitors do not choose a particular destination based on its cultural equity. However, they consume culture as a significant component of their visit. Bywater refers to them as *culturally motivated* visitors.
- The third type chooses a particular destination and travels to it because they want to feel the culture at a particular place. These are *culturally inspired* visitors.

Similarly, Stebbins (1996) classified cultural tourists as *general* and *specialised*. Generalists visit different cultural sites over time. They gradually increase their knowledge of different cultures. With time they may become specialists who repeatedly visit select cities, regions or countries to broaden their understanding of its culture. They may travel to the interiors of this destination in search of examples of culture in terms of art, music, festivals, and other things. They may visit museums for a deeper understanding of the culture.

Later, McKercher & du Cros (2003) came up with a hierarchical classification of cultural tourists based on the depth of the desired cultural experience of the visitor.

- At the bottom of the ladder is the *incidental* cultural tourist. Cultural tourism plays no role in the choice of destination, but once there, this type of tourist participates in some cultural activities with only a limited impact.
- The next level is the *serendipitous* cultural tourist. This type also does not travel for consuming culture. However, once they get to the destination, they engage actively with cultural products that profoundly impact them.
- For the *casual* cultural tourist, the cultural equity of the destination is a minor motivation for travel, but there are other more dominant motivations. As a result, they end up having a shallow cultural experience.
- At the second rung of the ladder are the *sightseeing* cultural tourists, who choose a destination for cultural visitations. However, they glance at the cultural products and return with a less deep cultural experience.
- At the top of the ladder is the *purposeful* cultural tourist. This type is like the *specialist* of Stebbins or the *inspired* visitor of Bywater. The prime motivation is to consume a cultural product, and the visitors engage in a profound cultural experience. They enjoy intellectually stimulating experiences, and often expand their understanding of culture by visiting museums, art galleries, temples and lesser-known heritage sites.

A similar typology classifies cultural tourists as *purposeful, tour-amateur, occasional, incidental* and *accidental* cultural tourists (Petroman, et al., 2013).

While walking tours of historical sites and monuments and heritage walks are vital instruments of cultural trips, in this chapter, we will focus on guiding tourists to archaeological sites, places of worship, and museums. We have already discussed walking tours and the principles remain the same while taking tourists through historical sites.

Guiding a group at an archaeological site

Archaeological sites and historical places are major tourist attractions the world over. An archaeological site is an identified area (or cluster of physical sites) where evidence of history is found and preserved. It could be related to prehistoric to relatively recent historical activities. Mohenjo-Daro and Harappa in Pakistan and Lothal and Rakhigarhi in India are examples of prehistoric civilisations in the Indian sub-continent. Given the antiquity of these places, the finds are fragile. These antiquities are non-renewable cultural resources and must be preserved. Archaeological sites throw up many challenges. Not all of them are easily accessible or interpreted without assistance. They are often in ruins and require a long restoration process, and some are still being evaluated by historians. They may exist in a natural ecological setting or be home to indigenous communities. Some vast archaeological sites are perpetually under excavation.

Archaeological sites pose two challenges. First, they are fragile assets, and as such, they must be visited with much care and sensitivity. The tour guide must mediate between the site, its ecosystem, its residents, if any, and the guest. Therefore, tour guides must be trained experts who can ensure that no harm is caused to the site while visitors are taken around. Many sites mandate that visitors can go around only under the supervision of trained and licensed guides (AMASR Rules, 1959).

The second challenge is that they are an ensemble of archaeological information distant from contemporary society. Therefore, an interpreter must contextualise it by establishing a relationship of artefacts discovered with their surroundings. They must also bring in the historical, geographical, sociological, anthropological, and other relevant contexts.

An archaeological site has two distinguishable elements: *artefacts* and *features*. Artefacts refer to movable physical objects like pottery, toys, flints and arrowheads, and others. Immovable elements of an archaeological site are referred to as features, for example the remains of hearths and houses. In addition, *eco-facts*, biological materials such as bones, scales, and others that have not been deliberately modified, may be present at many archaeological sites.

Concerned governments enact laws to preserve and protect archaeological sites. These laws prohibit the removal of any cultural (and in some cases, natural) materials. The tour guide must be aware of these rules and codes of conduct at an archaeological site. Massive tourist traffic harms the site, and inadequate site management results in deterioration or even destruction of the site and its related social, historical, educational and economic potential.

Preparing for the walk

Before leading a group of visitors to an archaeological site, tour guides must prepare beforehand and take necessary precautions so that no harm is caused to these fragile cultural resources. At the same time, they must ensure that guests are served a rich interpretation and a good time.

The first preparation is that the tour guides must be conversant with the local cultural and material history. All interpretations will be made within this context. Such a site relies more on fact and less on folklore. Another important thing is that the guides must make themselves aware of the sensibilities and sensitivities of the site. They must have a good idea of its fragility and its carrying capacity – how many visitors can the site accommodate at any given time? What size of the group can be taken around without harm? Which sections of the site are closed, and which ones are accessible to visitors? This assessment should have been made earlier, and the route planned accordingly. Many sites place restrictions on the number of daily visitors or permit a certain number of visitors during a time slot. This arrangement must be honoured at all times.

Tour guides should also know the challenges before the site regarding its preservation and conservation. They must meet the site authorities to understand their perspective and expectations from the visitors. This understanding will help tour guides ensure that their visitors have a minimum impact on the site. Tour guides must keep informed of the times the site will be open, and if it is not accessible on certain days or due to archaeological or related activity.

The tour guides should assess the demand that the visits to the site will place on guests. For example, is it suitable for differently-abled people, children, the elderly, and likewise? Tour guides should also be aware of the facilities available at sites and inform the guests – availability of food and if it can be taken inside the site; availability of water, restrooms, medical facilities, and other services.

Immediately before the start of the walking tour

The first thing that the tour guide should do while taking a group for a visit to an archaeological site is to tell the visitors about its importance and give a sense of its fragility. Visitors must be instructed to follow all guidelines prescribed by the site authorities and to keep to designated paths and observe all signage.

Most archaeological sites are open to the sky. Ensure that the guests have hats and sunglasses. If it is scorching and sunny, they must use sunscreen to protect their skin from prolonged exposure to the sun as they walk around the site. Visitors must carry sufficient water or energy drinks.

As you begin to walk

Again, as you begin to walk with the group, remind the guests what is permitted and what cannot be done. Tell them to keep to the main tracks and paths. It is easy to be lost in a big site. In that case, tell them how to navigate and get back. Guests should clearly see your concern for the site and observance of the appropriate behaviour.

On the tour

While visiting an archaeological site, the tour guides must remain vigilant and supervise client actions to ensure they are not harming the site. They must ensure that the guests remain on the marked pathways respecting all signage and fences. Sometimes there are ongoing or open archaeological excavations. Guest should avoid these sites. Ensure that the guests do not climb, sit or stand on archaeological structures or remains, or that they removing anything. If anything is picked up from the site, ensure that it is returned to the same spot before the group moves on. Remind them that removing objects from archaeological sites is illegal.

Littering an archaeological site is a sin. Visitors must use garbage bins and help in keeping the site clean. If tour guides come across any vandalism at the site, the same should be immediately reported to the authorities.

Guiding techniques

All the general techniques discussed earlier are applicable during the walk around the archaeological site. Interpretation is the key. All information is in bits and pieces and differs from the present-day context, and is not easy to comprehend. Guides should weave a story and paint a socio-cultural-historical picture of the artefacts and features. The tour guide may use panels, displays, renders, drawings and models to make a point and explain the concept. In addition, they may use a smartphone or tablet for sharing helpful information.

Finally, the tour guides must lead by example. Being sensitive to impacts and protecting and preserving the site are the most critical concerns of a tour guide at an archaeological site. In addition, they should be role models for good and responsible behaviour for the group.

Important precautions

Tour guides should never touch an artefact or lean against a structure or feature nor stand with their foot resting against the wall or a similarly placed feature. Ensure that visitors and others do not damage the monuments by touching them, leaning on them, sitting or climbing on them, even for the sake of photography. Writing on them is an absolute sin. Children should be discouraged from running around. Pointed things may cause harm to the features in a site. For example, pointed heels and studs and sharp walking sticks as they may cause damage to the sandstone floor and steps at the monument.

Excavated sites are very frail. Do not let the group walk on excavated sites, or even on their fringes, as it may damage the sections. Excavated material should never be displaced, sat on or removed. Ensure that visitors do not pick up any souvenirs/artefacts – even purchasing authentic objects is illegal in most cases. In some cases, children are engaged in selling activities at sites. Advise your group not to buy anything from children as it encourages them to drop out of school.

Photography is usually permitted. However, there may be some restrictions. Seek permission for photography. Photographic flash may cause damage to some paintings, frescos, artwork, and other exhibited elements. Seek permission to photograph the local community.

Archaeological sites usually do not allow pets. Ask the guest to leave their pet animals behind and not bring animals to the sites. Off-road and all-terrain vehicles, mountain bikes, and others, are not permitted in archaeological sites. Finally, the tour leaders must ensure that an archaeological site is not littered. Guests must not leave *anything* behind.

Knowledge bite

Tour guides should understand these terms.

- **Conservation:** Aims at prolonging the existence of artefacts and features. Conservation is about keeping objects and collections for future generations, as in museums
- **Preservation:** Aims at maintaining the status quo. It includes steps taken to minimise the rate of change and slow down further deterioration to maintain an object in its existing condition. Preventing objects from damage is part of preservation.
- **Restoration:** Aims at bringing back an object to its earlier known condition. It includes modifying the structure of an object and even reinforcing it with new material.

Guiding a group at a place of worship

In general, religious tourist sites can be divided into two: one, the places of worship like a Meenakshi temple in Madurai or Akshardham temple in New Delhi, a mosque like Al Aqsa in Jerusalem or Hagia Sophia in Istanbul, or a church like Notre Dame de Paris or Sagrada Familia in Barcelona; and two, the sacred places like Dargah of Khwaja Moin-Ud-din-Chisti at Ajmer, or Har-ki-paudi in Haridwar, or the Wailing Wall in Jerusalem. There are sensitivities associated with these places, making tourist guiding here a challenge. For example, in most religious places, the devotees are in some state of meditation, talking with and among the group is restricted. There could be other restrictions as well. An extreme example is that non-Parsis are not allowed to enter a Parsi Agiari (Fire temple). Similarly, non-Hindus are restricted from entering Jagannath temple (Odisa), Meenakshi Temple (Madurai) or Tirupathi Temple (Andhra Pradesh); and Duomo di Milano does not allow men with shorts and girls with short skirts. While most religious places allow visitors, they have a code of conduct that must be adhered to. It is always advisable to learn about the critical sensitivities of all the prominent religions: Hindu, Muslims, Christians, Buddhists, Jews, Jains, and especially those where an individual intends to guide tourists.

The believers of the faith may be interested in the history of the religious site and would like to spend some time; they would seek help with the protocols of the visit. For visitors from other faiths, one of the purposes of a tour to a religious place/sacred site is to educate themselves and witness a different culture or tradition. Such visits also help promote understanding, inter-faith tolerance and mutual respect. It may be noted that often silence needs to be maintained at religious sites. Therefore, there is little opportunity for narrations within these places. Tour guides may brief the group outside the premises, and advise them to explore the place independently.

The responsibility of a tour guide is thus threefold. One, to ensure that the worshippers are not disturbed by the visitors, two, to interpret the site and the religion for the visitors, and three, to advise and monitor the visitors so that they do not offend the religious sentiments of the faithful.

■ Preparing for the visit

A religious tourist place is usually considered important for its history, connection, place in the religion's tradition, and experience. It requires mediation by an expert who can hand-hold a visitor through their experience. Therefore, a tour guide must learn about its history and religious importance, and also learn the significant sensitivities, traditions and practices. For example, they must find out if there are specific requirements like dress codes, or covering the head.

The tour guide, to facilitate the visit, must gather information about prayer times or restricted access during certain rituals. Find if there are preferred or particular times for the guest visits (*darshans* in the case of Hindu temples). Tour guides should be aware of special events and festivities in religious places. They must also learn about the photography policy – is it allowed or not? Is it restricted in some areas of the place? Are any restrictions on eatables? Are there restrictions on any particular type of food? Can guests carry their mobile phones inside or not? How much will they have to walk or climb?

Many visitors would like to pay obeisance to the religious edifice by contributing. The tour guides must obtain information about donations and related policies. Also, locate utilities like – washrooms, lockers, parking, waiting areas, shops selling religious stuff, souvenir shops, and other services.

For interpretations, tour guides must learn about the structures' architectural details and what is unique to the location and the religion. They must tell what is done, where and what structural features support the activities. They must learn folklore about the place and religion. Usually, religious sites have a factual history since often they were made hundreds of years ago and then they have a popular belief associated with its inception, which is not necessarily factually accurate. A tour guide must know and educate their tourists with both.

Immediately before the visit

A religious place is primarily for the faithful. Believers have the first right, and their religious sensitivities are natural and must be respected at all times. Vvisitors have only been allowed access, but it is not a right. Visitors will often want to experience the place and not just see the religious structure. They would like to participate unnoticed without causing any distraction for the believers. They need the help and advice of the tour guide to facilitate this visit.

The tour guide must brief guests about the significant sensitivities, including silence zones, if any, and tell them about protocols to be followed. Tour guides must also ensure that guests enter as stipulated – cover heads, wash hands, take off shoes, and other details. Often there is a crowd in the religious sites. It is important to keep the group close and ensure that no one is lost. Make them aware of all the exit points and where they can collect if they have strayed away from the group. Ask them to take care of their belongings. In a crowded place, the chances of misplacing things are high.

Most religious sites are in places that require a lot of walking and climbing. Make sure that the guests are aware of that. Tell them how far they will have to walk and ask them if it is feasible for them to continue. If not, you can ask them to wait in an eatery, bus stop, or waiting area for the group.

As you begin the visit

Often there is limited opportunity for commentary inside the sanctum. Therefore, tour guides must offer a detailed briefing before the group enters the shrine. This briefing essentially has three components: the place and religion, architecture, and logistical issues.

The tour guide may begin by talking about the importance of the site within the context and history of the religion. After that, the guide may interpret the architectural plan of the site. Religious structures usually have specific geometries and orientation. Tell about areas/zones like sanctum sanctorum, altars, and other features. Also, tell about their importance or uses in religious rituals.

Finally, guides must brief the group about the time they may spend there. Since talking inside the shrine will be limited and the guests should not walk as a group because that could be distracting, they should disperse and tread individually. Towards the end of their visit, they should be able to find their way back. Tell them where to reassemble at the end of the visit and always tell them a fixed time they must come back. The tour guide must be present at that point ten minutes before the group arrives. Often associate it with a towering feature of the site – a tower, an important gate, a flag hoist, or anything similar.

On the tour

The tour guides must remain vigilant as the group visits the religious place. They must ensure that the protocols are observed. They must keep scanning the movements of their guests as far as possible. Since the group has dispersed and the guide can not be with everyone, they should remain alert and swiftly move towards them if anyone requires assistance. As and where they get a chance, they may explain religious objects and symbols, interpret them for the guests, and describe rituals, customs and religious practices (in a positive light). Build links to the lifestyle and community.

They must ensure that the guests maintain silence or a desired level of decibels when the guide or any guest talks and that the group's movements do not obstruct/hinder religious routines or people paying obeisance or performing a ritual. Understand that it is not a commodity; it is a place with sentiments attached. Therefore, interpret the site with emotions and sensitivity.

End of the visit

At the end of the visit to the religious site, ensure that all the group members are back. Advise the group on donations and where can donations be made. Also, inform about the purchase of souvenirs, if they can be made at religious places.

Important precautions

It is essential to be sensitive to 'do nots' in religious places. Inadvertently, the guests may hurt the sentiments of the followers of the religion. Religious places demand respect for the belief system. Therefore, advise guests not to touch, climb, and do anything contemptible to the deities or figurines in religious sites. If there are children, remind them not to do these. Never feed the animals at religious sites. Also, read and follow all the instructions mentioned on the site.

Devotees come to religious places to worship or to meditate. They do not like any distractions. Therefore, guests should avoid talking loudly and put their cell phones in silent mode. Public display of affection – holding hands, hugging, kissing, or any similar thing should be avoided as far as possible. Guests should dress appropriately by wearing clothing that covers their shoulders and legs. For women, they should avoid tight clothing. Why? It is a religious site, and they consider it improper by wearing clothing that 'reveals'.

Knowledge bite: Protocols for visiting religious sites

Each religious site has its protocol. A tour guide must be aware of these protocols and must remain sensitive to the concerns of these sacred places.

Hindu Temples

A Hindu temple is also called a *Prasadam*, *Mandir* or *Devasthanam*. Hindu tradition is a polytheist, so the temples are dedicated to one of the gods or a family of gods. The ideas of Hindu tradition and structures of their sacred places are rooted in Vedic traditions (Kramrisch, 1976). They usually deploy circular and squarish or rectangular patterns. It is considered a place where believers can meet their divinities.

A temple represents the Hindu cosmos – representing the elemental gods and the victory of good over evil. The statue of the chief deity of the temple – the *Moola Virat* – is placed in the inner sanctum, the *Garbhgriha*. Right above this is the central tower, the *Shikhara*. The corners of the temple can have smaller shrines dedicated to family gods of the chief deity. The entry is through a porch called *Ardhmandapa* or *Mukha Mandapa*. The murals depict the Hindu idea of cyclic time and the essence of life, and temples are often decorated with figures of gods and demi-gods depicting some philosophical concepts like *Dharma*, *Kama*, *Artha*, *Moksha*, among others. In addition, most temples offer a dainty called *Prasaad* or a liquid called *Panchamrit* or *Charanamrit* for the devotees.

The following may be noted while visiting a Hindu temple:

- Dress appropriately and do not reveal too much skin. Women should not wear sleeveless shirts and tops, while shorts and skirts should fall below the knee. Women should avoid wearing deep-neck blouses. Buttons must be fastened.

- Remain dignified and respectful and avoid loud conversation and laughter. Do not make comments criticising the religion or culture.
- Most Hindu temples may not allow any leather items into the temple, including belts and purses. Visitors should take off shoes and hats before entering the temples.

Mosques

A mosque is a romanised term for *Masjid*. It is a place for ritual prostration for Muslims. Muslims pray facing Mecca, and so the mosques have an ornamental niche in the wall, the *Mihrab* that indicates the *Qiblah* – the direction of holy Kaaba in Mecca. Mosques have a minaret, used for prayer calls (*Adhan* or *Azaan)*, five times each day. Believers must perform *Wudu* or *Wuzu*, the symbolic process of cleansing parts of the body. A mosque must therefore provide for a place for ablutions. There is often a pulpit used for Friday sermon called *Khutba*. Mosques have segregated spaces for men and women. Mats or carpets (*namaz*) cover the floor in rows for believers who bow and prostrate under the Imām's leadership.

In contrast to the Christian church, ceremonies and services related to birth and marriage are generally not performed in a mosque. Similarly, mosques do not provide unique structures for rituals like confession, repentance, and confirmation.

Visitors to the mosque may find a place to sit on the floor and take a moment for silent meditation or prayer. Afterwards, they are free to take a few photos if they want. However, one must not photograph worshippers and should not use a flash.

Essential things to keep in mind while visiting a mosque:

- Plan mosque visits when group prayers or sermons are not taking place. Avoid visiting the mosque during the half-hour following the prayer call.
- Step into the mosque with your right foot and leave with the left foot first.
- Do not disturb those who are praying, for example, by moving across them.
- The mosque is supposed to be a clean place. Remove footwear before entering, and leave it neatly outside the door alongside the other shoes. Wear unsoiled clothes. Offensive smells like those of onion, garlic, tobacco may be disliked. Similarly, one should not spit in the mosque.
- Do not conduct any buying/selling transactions in the mosque.
- One should remain calm and quiet in the mosque and avoid shouting and yelling, such as calling someone or for the lost items.

Christian Church

The church is a place for the believers of Christianity, and is used for religious activities and services. Often the plan view resembles a cruciform. Churches are typically oriented from east to west. The *narthex* or porch leading into a church is usually at the west end of the structure, while the chancel, sanctuary, and altar are in the eastern part. The broad central aisle in the western end is called the *nave*, and is where people usually sit during the service. Churches often have high domes or vaulted spaces to represent heaven. Most churches have a spire, a tall tower over the crossing or on the western end of the church. Churches follow the prevalent European architectural styles with local adaptations. As Christianity spread the world over, exciting history and folklore are associated with many churches. In addition, many churches have precious collections of murals and paintings.

Some of the crucial things to keep in mind while visiting a church include the following:

- Sunday and holy days are not the best time to visit with a group, and a church may not be open the entire day. Check for the time to visit. Try to avoid memorials and solemnities, and do not enter if there is a wedding, baptism or other event.
- Guests should be neatly dressed. Check for the dress code, if any. As in other religious places, it can be distasteful and even objectionable to show too much flesh. Ensure that the shoulders and thighs are not exposed. Remove headgear as a sign of respect.
- Guest should be mindful of worshippers and not distract them. They should not speak loud or make offensive or thoughtless comments. Cellphones should be in silent mode. Take care when walking through aisles as many Catholics kneel to worship.
- Do not make excessive displays of public affection.
- Keep your hands to yourself. Do not touch any artefacts, paintings or anything else that may have religious significance.

Buddhist Temples

Buddhism is one of the world's oldest organised religions, which mandated communal gatherings and required public and monastic spaces. A Buddhist temple, especially in India and South Asia, is essentially a cluster of three religious structures: a *vihara*, monastery; *stupas*, where venerated relics are housed; and a *chaitya-griha*, prayer hall. *Chaityas* are sometimes referred to as Buddhist temples. *Stupas* are the most characteristic structure in Buddhism. They are monuments rather than buildings, and commemorate a saintly person associated with that place. There are famous stand-alone *stupas* like Sanchi in Madhya Pradesh, India. In some cases like the *Mahabodhi* Temple at Bodh Gaya (India), the *stupa* is part of the *chaitya griha* and its central hall is designed to facilitate the circumambulation of the *stupa*.

Buddhist temples are places for quiet meditation. Any worship is done before the altar, lit with candles or oil lamps. Devotees burn incense and make offerings at the altar. The *chaitya-griha* has an open interior that houses cult figurines worshipped at the altar. The shrine also houses statues and paintings of Buddha, Boddhisattvas and Devas. The interiors, including the ceiling, are often painted, depicting tales conveying the Buddhist philosophy. The eaves and brackets are often decorated and colourfully painted.

Buddhist temples face south and sometimes east, but never north or west. Monks sometimes use the main hall of the *chaitya* to study and chant sutras. *Chaitya* may also have rows of boxed desks for keeping the scriptures.

Early Buddhist structures had an Indian origin. However, as Buddhism travelled eastwards, it incorporated the local architecture. For example, the 9th century Borobudur stupa in Central Java is the largest Buddhist temple in the world. It is like a stepped pyramid with a plan view of a stone mandala. Rock cut caves of South India are also important Buddhist religious structures. Some examples are the Ajanta and Ellora caves, and the Bhaja caves.

Buddhist temples are exciting places. Some of the things to be kept in mind are:

- There are religious services in a temple. If the guests arrive during that time, it might disturb the worshippers and meditators. Tour guides should check before they arrive.
- Often the faithful and believers meditate silently in the temple and may not be seen. Therefore, maintaining silence is essential. If required, speak very softly.
- Like any other religious place, one must dress modestly. Men and women should wear clean clothing that is not revealing, torn, skin-tight or low-cut, and not black.
- Advise your guests to refrain from public displays of affection.
- Visitors and devotees must remove their shoes before entering, but it is offensive to come barefoot, so guests must wear clean socks inside their sandals or sneakers.
- Visitors must bow their heads and respect the temple and the Buddha statues. Advise your guests not to turn their backs to the figurines, but walk backwards until some feet away. Circumambulations should always be in a clockwise direction.
- Advise your guests not to point at Buddha statues, monks, nuns or elders (especially not with their feet). Visitors should not touch Buddha statues, images, monks, nuns, and elders (especially on the head). A woman should never hand items to monks.

Sikh Gurdwaras

A *gurdwara* is the place for the religious congregation of Sikhs. Sikhism is a relatively young religion from north India with more than 25 million followers worldwide. *Gurdwara* literally means the doorway to the *guru*. Sikhs follow the teachings of ten human spiritual gurus whose preachings are compiled in the sacred text called *Shri Guru Granth Sahib*, also called *Adi Granth*, which is placed on a cot under a canopy in the gurdwaras. This is the most sacred part of a gurdwara. There are no candles, aroma, ritualistic bells, or other ceremonial apparatus in the gurdwara.

The *gurdwara* is central to the Sikh way of life. Besides being a place for a religious ceremony, Sikh children here learn about their faith, customs and traditions, texts and ethics. Adults connect to a Gurdwara for spiritual wisdom. Gurdwaras also offers food, shelter and companionship to those who need it.

Important gurdwaras host pilgrims during festivities. Most gurdwaras have a community dining hall called *langar*, where meals are served to the congregations. Langars are wholly managed by devotees, and in famous gurdwaras are tourist attractions.

Gurdwaras are open for everyone, from all faiths. Some important things to keep in mind while visiting a *gurdwara* include:

- Footwear are not allowed inside, to keep out dust and dirt. There is a special room for footwear in all gurudwaras. Ask your group to deposit their shoes there.
- One must wash his/her hands and feet before enting the *gurdwara*. This prevents the spreading of dirt and dust around the *gurdwara*.
- Everyone must cover their head. Many *gurdwaras* help with scarves to cover the head, which may be borrowed and deposited back on leaving the Gurdwara.
- When people enter a *gurdwara,* they must bow before *Guru Granth Sahib* touching the floor with their forehead.
- Inside the temple, everyone must sit on the floor. This act represents humility before the *Guru Granth Sahib* and conveys that everyone is equal and must sit together. Men and women must generally sit on separate sides of the hall.
- While sitting, one must not point their feet at the *Guru Granth Sahib*, considered disrespectful.
- Stand up and press your both hands together at the time of *Ardas* (the supplication).
- *Parikrama* (circumambulation) going around the *Guru Granth Sahib* or the *gurdwara* must be clockwise.
- Every *gurdwara* has a *langar* where anyone and everyone can have free meals. Many visitors are amazed by this idea and the management of these public kitchens.
- Alcohol, drugs, cigarettes or weapons are not allowed inside the *Gurdwaras*.

Synagogue

In simple terms, a synagogue is a building where Jews worship and study their religion. It is a community house of prayer that serves as a place for solemn ritualistic services and assembly and study.

Besides the traditional functions, modern synagogues permit social recreational philanthropic programmes. Synagogues are autonomous and democratic, established by the Jewish community who seek God through prayer and religious studies. There is no sacrifice in Jewish rituals, and no priests are required for common worship. Maintaining a temple and its rabbi is based on the local community's desires.

There is no prescribed architecture for a synagogue. However, there is a typically an ornamental *Torah Ark* where the scrolls of *Torah* are kept. The Ark is the holiest spot in a synagogue, and is located so that members of the congregation facing it, face Jerusalem.

There are two candelabra with candles constantly lit before the Ark. A raised platform called *Bimah* for service and reading out *Tanakh* or *Miqra*, the Hebrew bible. There are pews for members of a congregation. In orthodox synagogues, there are separate places for seating men and women. Some may house a *Mikvah*, which is a ritual bath.

Jewish synagogues are rather conservative. However, the following should be kept in mind while visiting a synagogue:

- All male visitors must wear a *Yarmulke*. This is a small, circular cap worn at the crown of the head. These are commonly available at the entrances of the synagogue. Women only need to wear hats or head cov erings if they are called upon to open the Ark holding the Torah. Men must be formally dressed. They may wear a suit, shirts, jackets with trousers while women wear suits or dresses that cover their upper arms.
- Everyone must stand when the congregation is asked to stand. Then, the rabbi tells people when to rise and be seated. Everyone should stay in the sanctuary whenever the Ark, the cabinet holding the Torah, is opened. One can leave to use the restroom or any other reason when the Ark is closed, and the rabbi says it is OK to sit down.
- One must keep prayer books on their lap or in the book-holders provided in seats. It is considered disrespectful to put holy books on the floor.
- Kiss anything holy that has accidentally fallen on the floor. These objects include *Yarmulkes* and prayer books.
- It is considered rude to attend calls during the prayer. Similarly, applause or chatting conspicuously with worshipers must be avoided.
- Guest may be advised against taking photos while inside a synagogue on Friday nights or Saturday mornings. This time is the Jewish Shabbat or Sabbath, and photography is prohibited during these times.

Leading at a museum

Museums are an authoritative intervention in educating visitors about the place, especially its history and legacy. People visit museums to make sense of what they see at the destination, including the historical and cultural sites. The exhibits weave stories to enthral the visitors and give them a sense of their time and place in history. Museums often complement the interpretation of a place.

A museum is an establishment that cares for and houses an extraordinary collection of scientific, artistic or historical artefacts. They can have their collections or show roving exhibits. Museums deploy curators, docents or volunteers to help their visitors interpret the collections. Many museums permit professional tour guides to lead visitors. The vision and objectives of a museum may vary, and so does their approach. Some favour education over conservation, or vice versa, but there is always some combination of both.

Museums can be either public or private. However, what differentiates a museum is the type of collections it showcases. These include paintings, drawings, photographs, sculptures, frescos, textures and textiles, prints, costumes, crafts, collages, artists' books, abstract installation, and many others.

Museum tours take several forms. They can focus on notable works, or on themes in the collection, such as families in art or still life painting across cultures, or look at a special exhibition open for only a limited time, or they can be a general walk-through of the collections to get visitors comfortable in making their own discoveries. Typically, tours involve walking with a group. Most museums nowadays are accessible to those who might be incapacitated.

Museums are restricted places that display selected artefacts. In many cases, the tour leader will seek the assistance of a local expert available in the museum – a docent or aide – to interpret cultural or historical information about the collection to visitors via an organised walk-through. The museum aide may take the group through the permanent collection or a special exhibition. In those cases, the tour leader will act as a sweep. In the absence of a museum aide, the tour leader may act as the guide to the group. He/ she must possess specialised and in-depth knowledge of displayed artefacts. It can be very intensive at times.

Preparing for the museum visit

Museums display selected exhibits that have deep meaning and specific interpretations. Tour guides must therefore spend time doing extensive research on the museum's permanent exhibits, and should prepare detailed notes for each exhibit (but not read them to the guests – spontaneity is the call of the hour).

It is not possible to interpret at length each exhibit. Instead, tour guides may prepare a couple of stories for different audiences and different duration of museum visits. For each storyline, identify the relevant exhibits and the time one would like to spend there. Some museums are famous for particular exhibits. Include them in your stories. For example, South Asians visiting the Jewel House in the Tower of London would like to see the 105-carat Koh-i-Noor diamond; anyone visiting the Salarjung Museum in Hyderabad would not miss the Veiled Rebecca by Italian sculptor Giovanni Maria Benzoni. Tour guides should memorise key dates and essential information and practice giving a tour.

If you are new to the museum, carry out a reconnaissance and visit the museum's website. Learn about any temporary exhibitions, recent redisplays, and the current installations. Learn the museum's policies regarding: opening and closing times, camera use, food and beverages, carrying a backpack/ hand baggage, and tickets and concessions, if any. As a tour guide, you must have information about the facilities in the museum, including the location of restrooms, cafeteria, souvenir shops, and others. Sometimes, as a tour leader, you may allow your guests to explore the museum independently. For those cases, identify meeting and re-meeting points.

Figure out facilities for guests with special needs, and talk to the tour operator to find out if there are any special requirements/needs of the group members.

Immediately before the museum tour

Tour guides must reaffirm rapport with the museum authorities on the day. Go and introduce yourself to museum staff. Request from them and arrange for sufficient material to be distributed among the group – maps, floor plans, brochures, and other things. Check for the bus drop off – this is generally the location where the tour guide meets the group if not accompanying them on the city tour.

A guide must always cooperate with officials and volunteers of the museum.

As you begin the visit

Start by introducing the museum to the group. Tell them subtly, at the beginning of a museum visit: "So here we are at....". Ensure that every guest has a valid ticket. If not, these must be purchased. Inside the museum, tell the guests about the washrooms, lockers, cafeteria, souvenir shops, and other facilities.

Identify the importance of the museum and anything special about it. After that, tell them the time that the group will spend in the museum. Also, indicate if you will allow any spare time to explore the museum independently.

If required, discuss the general layout of the museum so that the group members have an idea of where they are, how they will move and the point where the group collects after the visit.

Share with the group the essential things that they must bear in mind while traversing through the museum, including the policy for photography, use of mobile phones, and managing the backpack, if these are permitted inside.

On the tour

Visits to museums call for conscious attention to the following:

Handling objects and exhibits with care

Tour guides should be role models for the guests. They should not touch anything. They must continuously advise their group to refrain from touching objects. In some cases now, haptic augmented interactive technologies are deployed to allow visitors to make sense of the actual physical non-touchable artefact in an embodied way within a 3-D digital environment, using their sensory and motor skills.

Some exhibits are sensitive to a torch or even a laser pointer. Tour guides should check beforehand with museum staff if there are any directions to this effect. Advise guests accordingly.

Increasingly, museums are using interactive displays for non-personal interpretation. The tour guide can ask for a volunteer from the group to operate the buttons/touch the screen.

Remain sensitive to other visitors

Besides your group, usually, there are other visitors to the museum. A tour guide should ensure that there is always a pleasant and congenial environment in the museum. Visitors do not have to compete with others. As a role model, the tour guide should display respect for other guests and expect the same from his group. Remain aware of the other visitors. They are also there to enjoy. You and your group should not annoy them. Allow them space. Ensure that while your group moves within the halls and galleries, you occupy space in front of a display only when other visitors move out, and only while making the commentary – otherwise, you make space for others. Finally, ensure that your group does not block the alleys and pathways.

Be respectful and courteous to other visitors. Gently smile if you establish eye contact with others. Ensure that your group talks gently and appropriately, respecting the presence and privacy of others.

Take care of your guests

The tour guide should always be mindful of the guests' needs. Be proactive in reading and interpreting your guests' body language. This is critical. Some of your guests may genuinely seek detailed information, while others may be OK with a brief outline. Notice if your guests seem uneasy or gasping; you are probably boring them with too much information.

Frequently ask if anyone has questions for you. Appreciate the questions. Repeat them so that everyone in the group can hear you. Answer proactively.

Ensure that every group member has a fair opportunity to watch the display closely if that is required.

Guiding techniques for the museum

Most of the techniques like position triangles and TVPs discussed earlier are applicable while guiding in a museum. However, a museum is often an enclosed space, and that factor may be considered. These include the following.

Positioning

Never start talking unless everyone has gathered and are ready to hear from you. Ask your clients to be in a close group around you. Before starting to speak, ensure that everyone is there and comfortably situated.

The principle of the position triangle applies here as well. Remember a triangular formation so that the group can see the object and, at will, can observe you speak also. The triangle with a 120° spread works particularly well in museums. Ensure that you have eye contact with the audience. You would usually talk to the group but intermittently glance at the object. Never speak facing the object to avoid eye contact with the group.

Be precise in pointing out objects, for example: 'the first figure from left on the panel, with three heads, is Lord *Brahma*...'. For a small object, allow everyone to turns to come forward and see the object from close. You may help individual members if required. Then, point it out again at the end and allow everyone a chance to glance before you leave that spot.

Often, there will be a situation when the display is at a lower height and allows visitors to go around it. In this case, the tour guide may get behind the object while the group stands in front of it. It is similar to the typical classroom layout where the teacher explains an experiment on her desk while the students watch it from the other side.

Building links

Often, the museum relates to the history and legacy of the place. For example, it houses artefacts found in ruins of a palace or fort or an excavation site. Alternatively, it has objects that were part of some determining phase or event of the history of that place. Connect or build links from the artefact before you with what the group might have seen elsewhere in the city or monuments or relate to something from the culture or history.

Let the objects speak. Identify artefacts to stop by and include in your commentary. You should have interesting stories about these to share with guests.

Another valuable and exciting way is to allow the group to express its opinions, which you may subsequently moderate or rectify. You may try to provoke a response: 'last time I was here, the guest said.... what do you think?' This technique will ensure active participation of the visitors, as against them being passive listeners of the narration.

A tour guide must be ingenious, creative and inspiring. Museums have their logical sequence of displaying the exhibits. However, a tour guide can weave exciting stories, links to objects in different sections and rooms, and even engage guests into spot quizzes or games. Many museums have also developed games and activities for visitors, especially children.

Movement in the museum

Museums being enclosed spaces with select artefacts on display merit attention in so far the movement of visitors is concerned. It may be noted that visitors typically move more leisurely in a museum compared to a walking tour in the open. However, the group may move from one exhibition hall to another, and pass through passages and open spaces in between. A tour guide should keep the group together. Count them time and again to ensure everyone is there. If you are a tour leader and are not leading the group within the museum, you may act as a sweep.

If the group needs to walk some distance to another room or hall, announce clearly where the group need to get to next: 'please follow me to the room across the corridor, we have the armoury displayed there.'

Call the group's attention before moving to the next stop of interest. Then, locate yourself vis-a-vis the exhibit and allow the group to assemble at the third vertex of the position triangle spreading across 120°.

Sometimes, the group might encounter things that will attract their attention but not fit your theme. Acknowledge these Top Visual Priorities (TVPs) and quickly return to the point that you were making.

Handling tagalongs

Another significant challenge that a tour guide may face while leading in a museum is people who might join your group without realising it is a private group. These are called 'Tagalongs'. Tour guides are expected to tactfully (maybe indirectly) tell them that this is a private group – 'my group, as I told you yesterday...' or, for example, 'you will see a similar figurine when we get to the palace in the evening...'. This call indirectly communicates that you are a private group and that you are leading the same.

If the tagalong fails to receive the message and continues with the group, the tour guide should find a suitable opportunity to tell them directly that this is a private group. Advise them that there are volunteer /aide facilities in the museum. They may engage them separately.

The sign of professionalism is that tagalongs are handled subtly without the visitors noticing, and without causing any obstruction, breaking momentum or creating bitterness. Never be rude to anyone, including tagalongs.

End of the tour

As the visit to the museum comes to an end, the guide must announce the conclusion of the visit. Then, help the group with information about conveniences like washrooms, souvenir shops and cafeteria. Thank your museum guide, if one has been used. If the tour guide assignment was restricted only to the museum visit, you would say goodbye to the group now. If it was a part of the more extensive itinerary, inform the group about the next moves. If required, lead them to the stop for getting back to the coach.

Please note

Exhibits in the museum have been removed from their original locations (and therefore contexts) and preserved and presented. It is, therefore, crucial that these objects are correctly interpreted. According to *Interpretation Handbook and Standards* (Colquhoun, 2005),

> "Interpretation communicates what is significant about places, people or events. The essence of interpretation provides insight for visitors about what is special and how and why it is valued. It is a celebration of place and things, culture and nature, creativity and folklore, great successes and failures in our history."

A tour guide's job is to bring back to life these objects.

Interpretation of artefacts – a suggestion

A tour guide may follow a four-step explanation of the exhibit.

1. Begin with two fundamental questions:
   - What is it?
   - What is its importance?
2. Next, explain what is visible
   - What material is it made? How was it made?
     - ☐ Talk about material and technology. Connect the technology to antiquity.
   - How did it get to the present shape?
     - ☐ Is it in its original shape? Was it fragmented or restored/rebuilt at some time?
   - Talk about the environmental context: the climate, local materials, etc.
   - Talk about the socio-cultural context: try to connect it to contemporary lifestyle, tradition, culture, and other dimensions.
3. Then, bring in the invisible aspects of the artefact
   - What was the purpose of the object?
     - ☐ Who built it, and why?
   - Where, when in what numbers?
   - How difficult was it to make/build it then?
4. Finally, try to broaden the interpretation by building links
   - Compare the object and its times with other things elsewhere in the world (better if it is the guests' context)
   - Bring in ethnological, archaeological, historical, geographical, geological, and other dimensions.

Important precautions

Several things must be considered during a visit to a museum. As a role model, the tour guide must never touch any exhibit as they tell about them. They must remain considerate of other guests in that they do not block their movements, and ensure that the group does not do so. Tour guides are not omniscient. Sometimes you do not know the answer to a question. Do not cook up a reply. Do not present unsupported details as if they were facts. However, share a popular belief, but with a disclaimer that it is just a belief.

The attraction of nature

Human survival depends upon the surrounding ecosystem. The connection between humans and nature is perpetual, and both are inseparable. However, the concept of anthropocentrism-human chauvinism, the viewpoint that highlights human supremacy over nature, has created problems in nature and disturbed the ecological balance (Grey, 1993). The conflicts are visible in climate change and the increasing recurrence of natural calamities. First, Homo sapiens, for safety from animals and wildlife, distanced themselves from nature. Later, with the motivation to secure a comfortable lifestyle, the distance grew broader as the built environment shaped our cities.

The advancement of technology further separated humans from nature (Schultz, 2002). Interestingly, the Biophilia Hypothesis (Kellert & Wilson, 1993) suggests that humans are biologically dependent and attracted towards nature. Accordingly, humans develop an inherent desire to affiliate to nature and consider themselves part of nature. Several authors have discussed human-nature connection that revolves around humans being in touch with, connected to, or part of nature. Thus there is an apparent oneness and interconnectedness with nature (Schultz, 2002).

Rapid urbanisation emerged as the root cause of the separation between humans and nature (Chalquist, 2009). However, even after securing luxuries and comfort, there is some missing component that people explore in the natural environment. Those natural areas pull humans towards them. The terms 'urban dwellers' or 'city dwellers' take centre stage in this case. Perhaps the significant motivation of urban escape for dwelling in natural experiencescapes is a growing phenomenon of nature-based tourism compared to routine life (Fossgard & Fredman, 2019; Mehmetoglu, 2007). People prefer to spend time for short intervals for recreation, outdoor pursuits and relaxation in the natural environment because their significant portion of time is spent in the built environment (Fossgard & Fredman, 2019; Schultz, 2002) Travelling to nature and establishing a connection with it has underlining principles that have cognitive, affective, and behavioural components and lead to a psychological connection between humans and nature (Schultz, 2002). Holidaying in nature or indulging in nature based-tourism is based on three types of relationships between humans and nature: one, experiences dependent upon natural settings; two, experiences enhanced by natural settings; and three, experiences where the natural environment plays a subordinate role (Valentine, 1992).

To understand the dynamics of nature-based tourism and the motivations of nature-based tourists, we can interpret tourist experiences in two interchangeably used domains of tourism, nature-based tourism and ecotourism. Interestingly, both the domains of tourism appear similar, but fundamentally conceptual understanding underlines their difference. While any form of tourism that depends on the natural environment is nature-based tourism, ecotourism is concerned with tourists' responsible behaviour that considers the preservation of natural resources and welfare of the local community (Nyaupane, 2007). To understand in broader terms, nature-based tourists are attracted to nature either to develop a deep connection with it and understand its significance for educative purposes, or for recreation or relaxation concerned with multidimensional well-being. Simplistically, pursuits in nature-based tourism are directly connected with two types of crisis: the human and the environmental crises. To combat these, the need is to ignore nature's scientific observation as an object and develop an everlasting bond with nature (Tam, 2013). The need for time can be observed in Leopold's (1949) words, "We abuse land because we regard it as a commodity belonging to us. When we see land as a community to which we belong, we may begin to use it with love and respect." Leopold's approach brings some relief to both forms of crisis. So, the environmental crisis can be taken care of through ecotourism, but how can the internal crisis prevailing and growing in humans, particularly in urban and city people, be taken care of? Answering this brings us back to what pulls us to nature and what we are exploring in the natural environment?

Previous research work argued that reconnecting people is beneficial for environmental preservation and people's well-being (Beringer & Martin, 2003). People prefer to travel to initiate transformation and enhance multidimensional well-being (Little, 2012; Reisinger, 2013; Ross, 2010). Researchers are reporting an increasing number of tourists to national parks and protected areas. Involvement in adventure-based activities in natural areas such as hiking, trekking, mountaineering and many more have become a growing phenomenon (Kim et al., 2015). Cohen (1979) suggests that people use travelling as a medium to escape from harsh realities and the growing pressure of everyday life. In the quest to re-create the damaged identities and combat the existential crisis, people consider travelling as a tool to reconnect to self and then to the world (Maoz, 2006; Robledo & Batle, 2017). They escape from artificial urban setups and seek peace and tranquillity in natural and rural landscapes, usually isolated or in the wilderness (Ashley, 2007; Little, 2012). Nature is a source of awe and elevates humans spiritually by producing healing effects (Beringer & Martin, 2003). The movement of people towards nature provides them peace, solitude, self-engagement, meaning-making, tranquillity and thus, involvement in nature is therapeutic and navigates a way to spiritual well-being (Brown, 2006; Jepson & Sharpley, 2015; Owton, 2013; Sharpley & Jepson, 2011; Shin et al., 2012; Stringer & McAvoy, 2013).

The next question arises about the creation of nature-based experiences. How nature-based tourism experience is created and how it is different from other forms of tourism experiences. What are the requisites of enhancing nature-based tourism experiences? In general tourism, experiences are centred around enjoyment, engagement, and memory creation. In nature-based tourism, nature's experience is the product and value and not the end of the consumption process. Commercialisation becomes necessary when nature-based tourism becomes part of the experience economy (Pine & Gilmore, 2011). The value addition to nature-based experience is the focus of commercialisation and must be introduced by an agent. In nature-based tourism, these agents are presenters or guides who interpret, guide, co-create or facilitate nature-based experiences (Vespestad & Lindberg, 2011). Specifically, presenter involvement becomes significant in natural sightseeing, usually in protected areas or national parks. For example, a presenter's value addition can be made by highlighting the potential of spirituality in nature. To understand better, it could be underlining how nature and species enhance the spirit and emotions of participants (Vespestad & Lindberg, 2011). In nature-based tourism, a guide or presenter's role significantly changes as guiding takes place in an uncontrollable environment, different from a controlled built environment. In this way, the experiences are heavily dependent upon tourists, and therefore, presenters of nature-based tourism depend upon the assumption of a tourist. How a tourist seeks nature-based experience and how the same can be appreciated. Guided tours in the natural environment create experiences that may facilitate or improve well-being. An example can be the Shinrin yoku-forest bathing technique in Japan facilitated by guided tour providers from the Association of Nature and Forest Therapy Guides tours in national parks (Sheldon, 2020). Some tribes settled in the interiors of Hawaii, the USA, or India can also shed light upon the importance of nature, wildlife, and medicinal use, among other dimensions, through guided tours or displays of singing and dancing rituals.

Nature tours

Nature tours are discretionary travel to natural areas. Nature tourists are visitors who spend their time and money enjoying and appreciating a broad range of outdoor activities. This activity generates economic benefits for the local community on the one hand, while on the other, it encourages the conservation of ecological, social, and cultural values. Nature tourists are primarily urban, tired of stress and traffic, and want to get away. They are therefore not used to the wilderness and its ways. A nature guide's job is to help them adapt unobtrusively and in an undisruptive way as they enjoy nature and educate themselves. The nature guide must assist them at every step, almost handhold them through their tour. Nature tourists come with different motivations: appreciating pure nature, education and history, relaxation, social, weekend (hangout) travel, camping, etc.

There is rising awareness and concern about conserving and preserving the natural habitat as more people travel there. This type of travel is expected to increase in a post-COVID world. Nature guides are often locals. Besides having the necessary knowledge about the landscape and nature they are part of, they are also the custodians of their guests' responsible behaviour. They may need quite a technical expertise. They must never forget and ensure that nature has the right of way every time.

■ Preparing for the nature-walk

A tour guide must begin with researching and documenting relevant details about flora and fauna, culture and heritage of the area, such as a park, a sanctuary, or a forest, a village, etc.. Next, they must familiarise themselves with the customs, protocols, and rules, if any, of the area they are taking their visitors. Finally, they must be fully cognizant of the local sensitivities.

They must carefully plan the itinerary, keeping in mind the groups' ability and interests, distance, stops, gaze points, and other things. They must equip and clothe themselves appropriately for the trip they are leading. They must generally carry mosquito repellent, sunscreen, antiseptic cream, hat, socks and shoes. Concerned authorities must be informed about the tour, route, size of the party, and the expected arrival time. It is also essential to have an exigency plan to evacuate the guests in case of inclement weather, tiredness and other unexpected happenings.

■ Immediately before the start of the nature walk

Before setting out on tour, it is advised to check the latest weather forecasts, and inform the guests well in advance if need to carry anything in particular.

Introduce guests to the park by providing a general overview of its geography, geology, flora, fauna and communities. Give them an overview of the tour plan and activities, including fun and games. Educating the guests is very important, so tell them about sensitivities, dos and don'ts. Ensure that they carry hats, socks, shoes and warm clothing in winters. Visitors should preferably wear khaki, brown and olive green clothes when visiting a park or a protected area, as animals might get scared when they see bright colours as they are not used to them.

Ask the guests to take a minimum of disposable items with them and not to leave behind mosquito repellent, antiseptic cream, sunscreen. Ask if they have brought their binoculars, camera, phones, water bottle, and other things. All vehicles must be left outside the reserves and parks. Natural areas have their serenity, and this should be maintained by keeping out outside influences like music equipment and honking. Loud conversations are taboo in a nature park.

As the walk begins

Keep the guests on the trail and tell them not to stray alone in the forest area unless they are in a free access area. Closely observe your visitors as they spend the first few minutes or hours (as appropriate) and help them modify their conduct to conforms to acceptable behaviour in the natural area. Instruct guests to pay heed to all warning signs and stay away from restricted areas for their safety.

On the tour

Natural areas are often fragile ecosystems, and appropriate conduct is expected. Encourage the use of waste receptacles. Advise guests to use designated toilets, if possible. Otherwise, suggest burying human waste in 6-8 inches deep pits, away from streams and waterways. Guests must also be advised to dispose of cooking waste and dirty water away from water sources and campsites. The places should never be altered. Advise them not to trench around tents and to desist from building temporary structures, such as benches, shacks or similar structures. Guests may be allowed to swim only in approved areas. Most parks and reserves have guidelines for bonfires, and these must be adhered to. When driving through forests or other landscapes, the vehicles should remain on the designated roads and tracks. Off-road motoring is not advised as it causes damage to nature, and it might be risky.

Tour parties should avoid littering at all times. The use of paper should be minimised. Water should be used sparingly, and no detergents.

The ecosystem's three crucial constituents are plants, animals, and the local community. Therefore, necessary care must be taken to deal with each of these.

Advise the group not to collect plants or pluck flowers, nor damage trees or alter their habitat. They may not collect plants, animals, rocks, artefacts or firewood – leave things as they found them. In some cases, it is illegal and might attract prosecution. Guests should be refrain from graffiti on any surface.

Similarly, the guests should not approach or threaten the wildlife. They should be advised against feeding wild animals and birds, and especially avoid feeding animals junk food. It will change their eating habits which are not advisable when they have plenty of natural food in the forest. Guests should respect nature – keep noise to a minimum and be kind to creatures. They must desist from using strong-smelling substances like perfumes or deodorants, and must not carry food with a strong aroma into the forest. Animals are sensitive to odour, and any strange smells may arouse their curiosity, causing erratic behaviour. Animals tend to flee when they see a human being. Advise the group not to incite them out by throwing stones or other objects. They must keep a reasonable distance from all the animals and leave them in peace.

The third dimension is the local community. Educate guests to respect local traditions, customs, traditions and religions, and their privacy. Guests must seek explicit permission when photographing local people, especially their homes. Also, the guests should be advised to wear appropriate clothes.

Guiding techniques

Leading on natural trails can be organised in different formats depending upon what is permitted and desired by the tourists. Depending on where you choose to pursue a tour leader career, you may need to acquire additional skills like mountain guiding, difficult terrain guiding, survival skills, bush survival, life-saving techniques, etc.

End of the tour

As the trip ends, guests must clean up the place after the stay or activity and make it look like no one had been there. Guests must not litter the place with polyethene bags, tetra packs and other similar things. Litter spoils the beauty of the area, and can be harmful to the animals. Moreover, if some guests smoke and are allowed to do so in public on that site, they should be asked to save cigarette butts and take them back. Guests should carry all such garbage back with them to be disposed of properly at predetermined collection points.

End-of-the-tour is also the occasion to request guests to report their positive or negative observations to the authorities. The guide must own the responsibility to ensure that the place is unharmed and used appropriately. For example, they must report errant drivers or guests violating the park's rules. Any sightings of poachers should also be reported to the authorities.

■ Some other precautions on nature tours

Besides the above guidelines, the following must also be observed. Guests may not be allowed to collect wood, bone, horn, shells, artefacts or other natural and man-made objects. However, they may stop, touch and enjoy them. Then they leave them for others to observe and enjoy.

Visitors should refrain from using flashguns when photographing animals, and must not bring firearms into the forest. In some cases, this is illegal.

Key points

Cultural tourism is a popular reason for travel. Visitors experience the contemporary, to make sense of the place by visiting monuments, archaeological sites, museums, and places of religious significance. During a limited period of visits to the destinations, visitors often need experts to interpret these elements.

There are different ways of understanding cultural tourists. One way is to look at the typology: *interested, motivated,* and *inspired* visitors. Another is to look at them as *general* and *specialised* cultural tourists. A third way is to understand them as *incidental, serendipitous, casual, sightseeing* or *purposeful* cultural tourists. Finally, another typology classifies cultural tourists as *purposeful, tour-amateur, occasional, incidental,* and *accidental* cultural tourists.

Archaeological sites and historical places are major tourist attractions the world over. Archaeological sites pose two challenges. First, they are fragile assets, and as such, they must be visited with much care and sensitivity. The second challenge is that they are just an ensemble of archaeological information distant from contemporary society. Therefore, an interpreter must contextualise it by establishing a relationship of artefacts discovered with their surroundings. They must also bring in the historical, geographical, sociological, anthropological, and other relevant contexts. Guides must consider these two concerns to design and execute an interpretive plan for an archaeological site.

Visitors to religious sites may have different motivations. For example, they may want to visit purely to peep into the community's religious beliefs, rituals, and lifestyles. Alternatively, they might like to visit these sites because they often have a fascinating historical legacy. Finally, another section might be visiting the site because of their faith and spiritual connection. In any case, a site is a place of worship for believers who have the first right to be there and have a spiritual connection. The responsibility of a tour guide is thus threefold. One, to ensure that the worshippers are not disturbed by the visitors, and two, to interpret the site and the religion for the visitors, and three, to advise and monitor the visitors so that they do not offend the religious sentiments of the faithful.

Guides must be aware of the architectural details, the iconography of the religious structures, and the nuances of the religions associated with these sites. They must also be aware of tenets of the world's major religions: Christianity, Islam, Judaism, Hinduism, Buddhism, Jainism, Sikhism, and others.

Museums educate visitors about the place, especially its history and legacy. People visit museums to make sense of what they see at the destination, including the historical and cultural sites. The exhibits weave stories to enthuse the visitors and give them a sense of their time and place in history. Guiding in a museum is a specialised skill, and tour guides must possess these skills.

For a variety of reasons, humans like to travel to natural settings. Some thinkers believe that humans have an inherent desire to affiliate to nature, but people also participate in hiking, mountaineering and others activities in these natural settings.

Nature is a source of awe and elevates humans spiritually by producing healing effects. It provides people peace, solitude, self-engagement, meaning-making, tranquillity, and thus, involvement in nature is therapeutic and navigates a way to spiritual well-being. Visitors use travelling as a medium to escape from harsh realities and the growing pressure of everyday life. However, such travel has an impact on nature, including the communities. Guided tours in the natural environment create experiences that may facilitate or improve well-being, but they must ensure that the negative impacts are minimised.

Several concepts are associated with travel to natural settings. These include nature-based travel, ecotourism, sustainable, responsible, and green tourism. Often these are used interchangeably, but a professional tour leader needs to appreciate the differences and relationships between these varied concepts.

The chapter discusses the preparation for leading a nature walk, and the precautions to be taken. Nature guides may need to acquire additional specialist skills appropriate to demands of the area.

Key terms

Archaeological site	An area or physical site that provide information and materials concerning the distant past by preserving the physical remains. Usually, it is a part of an archaeological record that is or may be investigated by archaeologists.
Chaitya Griha (or chaitya)	In Indian religious tradition, a shrine, sanctuary, temple or prayer hall. The term is common in Buddhism. Its open interior houses cult figurines worshipped at the altar, and statues and paintings of Buddha, Boddhisattvas and Devas.
Church	A building used for Christian religious services and activities. The plan view of a church often forms a Christian cross.

Cultural tourism	A type of tourism activity in which the visitor's essential motivation is to learn, discover, experience and consume the tangible and intangible cultural attractions/products in a tourism destination. These attractions or products relate to a set of distinctive material, intellectual, spiritual and emotional features of a society that encompasses arts and architecture, historical and cultural heritage, culinary heritage, literature, music, creative industries and the living cultures with their lifestyles, value systems, beliefs and traditions
Ecotourism	A sustainable form of natural resource-based tourism that focuses primarily on experiencing and learning about nature and is ethically managed to be low-impact, non-consumptive, and locally oriented. It typically occurs in natural areas and should contribute to the conservation or preservation of such areas.
Garbhgriha	Refers to the altar room of a Hindu temple where the temple's principal deity's idol is installed and worshipped. Though the term is mainly associated with Hindu temples, it is also found in Jain and Buddhist temples.
Gurudwara	The doorway to the *guru*, is the Sikh place of learning and worship where the community gathers for congregational prayers, festivities, and activities. Visitors of any background can seek shelter, comfort, and food.
Langar	The community kitchen of a Sikh temple, the gurudwara. It serves vegetarian meals to all free of charge, regardless of religion, caste, gender, economic status, or ethnicity. People sit on the floor and eat together, and the kitchen is maintained and serviced by Sikh community volunteers.
Masjid	The Arabic word for a mosque, a house or open area for ritual prostration, especially the congregational prayers for Muslims. Though masjids are buildings with many architectural features, the structure is essentially an open space, generally roofed over, containing a *mihrāb* and a *minbar*, with a minaret sometimes attached to it.
Nature walk (nature trail/ educational trail)	A walk with an experienced guide on a specially developed hiking trail or footpath that runs through the countryside, along which there are marked stations or stops next to points of natural or cultural interest.
Nature-based tourism	Tourism to natural areas for various attractions or activities like birdwatching, photography, stargazing, camping, hiking, hunting, fishing, and visiting parks.
Parikrama	The ritual of moving clockwise around an object of devotion to indicate reverence in Hindu, Jain or Buddhist contexts. Parikrama means "the path surrounding something" in Sanskrit. It is also the circulatory stone pathway around the shrine of the temples in Hinduism, Buddhism and Jainism.
Stupa	The most characteristic structure in Buddhist architecture. A commemorative monument usually housing sacred relics associated with the Buddha or other saintly persons. Some believe that it predates Buddhism.
Sustainable tourism	The environmental, economic, and socio-cultural aspects of tourism development, and a suitable balance must be established between these three dimensions to guarantee its long-term sustainability.
Synagogue	A Jewish house of worship that serves as a place for liturgical services, assembly, and study. The main sanctuary is for prayers. There are rooms for study, a social hall, and offices. Some have a separate room for the study of the Torah.

Discussion questions

1. How is a cultural tourist different from other visitors?
2. Define cultural tourism. List some elements of your place's life that could be presented to visitors as cultural tourism products.
3. Visit an archaeological site near you. Talk to the site authorities to learn about the dos and don'ts for the visitors. Assess if these are good enough to prevent any harm to the site. What facilities can be offered to the visitors to make their visit enjoyable at that site?
4. What is the difference between conservation, preservation, and restoration concerning an archaeological site?
5. Visit a popular religious/ spiritual site near you. Talk to the chief priest or administrator to evaluate the visitors' opportunities to this site. List the concerns, apprehensions and sensitivities identified during this discussion.
6. Prepare an interpretation plan for the architectural details of a religious site near you. Identify the key stops and prepare scripts of the narrations at each.
7. Does your place have a museum? Visit one. What is the focus or theme of this museum? What are the famous artefacts on display in this museum?
8. Discuss how you would include the local museum in the broader interpretation of the place.

Activity

1. Visit the nearby natural areas. Then, meet the authorities to understand the restrictions they impose on visitors.
2. In a small group, enact a sketch to ask the tagalong to keep off the group.
3. Visit the local museum. Select a painting or a statuette. Then, research and prepare its iconographic interpretation.
4. Develop an interpretation for the local nature trail.

References

AMASR Rules. (1959). *The Ancient Monuments and Archaeological Sites and Remains Rules, 1959.* Retrieved November 6, 2020, from Archaeological Survey of India: http://www.asichennai.gov.in/downloads/amasr_rules_1959.pdf

Ashley, P. (2007). Toward an understanding and definition of Wilderness Spirituality. *Australian Geographer*, 38(1), 53–69. https://doi.org/10.1080/00049180601175865

Beringer, A., & Martin, P. (2003). On adventure therapy and the natural worlds: Respecting nature's healing. *Journal of Adventure Education & Outdoor Learning*, 3(1), 29–39. https://doi.org/10.1080/14729670385200221

Brown, G. (2006), Mapping landscape values and development preferences: a method for tourism and residential development planning. *International Journal of Tourism Research*, 8: 101-113. https://doi.org/10.1002/jtr.562

Bywater, M. (1993). The market for cultural tourism in Europe. *Travel and Tourism Analyst,* 6, 30-46.

Chalquist, C. (2009). A look at the ecotherapy research evidence. *Ecopsychology*, 1(2), 64–74. https://doi.org/10.1089/eco.2009.0003

Cohen, E. (1979). A phenomenology of tourist experiences. *Sociology*, 13(2), 179–201. https://doi.org/10.1177/003803857901300203

Colquhoun, F. (Ed.) (2005). *Interpretation Handbook.* Department of Conservation Retrieved November 12, 2020, from https://www.doc.govt.nz/globalassets/documents/about-doc/role/policies-and-plans/interpretation-handbook-complete.pdf

Fossgard, K., & Fredman, P. (2019). Dimensions in the nature-based tourism experiencescape: An explorative analysis. *Journal of Outdoor Recreation and Tourism*, 28. https://doi.org/10.1016/j.jort.2019.04.001

Grey, W. (1993). Anthropocentrism and deep ecology. *Australasian Journal of Philosophy, 71*(4), 463–475. https://doi.org/10.1080/00048409312345442

Jepson, D., & Sharpley, R. (2015). More than sense of place? Exploring the emotional dimension of rural tourism experiences. *Journal of Sustainable Tourism*, 23(8–9), 1157–1178. https://doi.org/10.1080/09669582.2014.953543

Kellert, S. R., & Wilson, E. O. (1993). *Biophilia Hypothesis*. Island Press/Shearwater Books.

Kim, H., Lee, S., Uysal, M., Kim, J., & Ahn, K. (2015). Nature-based tourism: Motivation and subjective well-being. *Journal of Travel and Tourism Marketing*, 32, S76–S96. https://doi.org/10.1080/10548408.2014.997958

Kramrisch, S. (1976). *The Hindu Temple*. Delhi: Motilal Banarsidass Publishing House.

Leopold, A. (1949). *A Sand County Almanac and Sketches Here and There.* Oxford University Press, New York.

Little, J. (2012). Transformational tourism, nature and well-being: new perspectives on fitness and the body. *Sociologia Ruralis*, 52(3), 257–271. https://doi.org/10.1111/j.1467-9523.2012.00566.x

Maoz, D. (2006). Erikson on the tour. *Tourism Recreation Research*, 31(3), 55–63. https://doi.org/10.1080/02508281.2006.11081505

McKercher, B., & du Cros, H. (2003). Testing a cultural tourism typology. *International Journal of Tourism Research,* 5, 45-58.

Mehmetoglu, M. (2007). Nature-based tourism: A contrast to everyday life. *Journal of Ecotourism*, 6(2), 111–126. https://doi.org/10.2167/joe168.0

Nyaupane, G. P. (2007). Ecotourism versus nature-based tourism: Do tourists really know the difference? *Anatolia*, 18(1), 161–165. doi.org/10.1080/13032917.2007.9687044

OCED. (2009). *The Impact of Culture on Tourism.* Paris: OCED.

Owton, H. (2013). Ecotherapy: Healing with nature in mind. *European Journal of Psychotherapy & Counselling,* 15(3), 300–302. doi.org/10.1080/13642537.2013.814239

Petroman, I., Petroman, C., Marin, D., Ciolac, R., Văduva, L., & Pandur , I. (2013). Types of cultural tourism. *Scientific Papers: Animal Science and Biotechnologies,* 46(1), 385-388.

Pine, B. J., & Gilmore, J. H. (2011). The experience economy. Harvard Business Review Press. https://doi.org/10.1007/bf03036337

Reisinger, Y. (2013). Transformation and transformational learning theory. In *Transformational Tourism: Tourist Perspectives.* doi.org/10.1079/9781780642093.0017

Robledo, M. A., & Batle, J. (2017). Transformational tourism as a hero's journey. *Current Issues in Tourism,* 20(16), 1736–1748. https://doi.org/10.1080/13683500.2015.1054270

Ross, S. L. (2010). Transformative travel: An enjoyable way to foster radical change. *ReVision,* 32(1), 54–61. https://doi.org/10.4298/revn.32.1.54-62

Schultz, P. W. (2002). Inclusion with nature: The psychology of human-nature relations. *Psychology of Sustainable Development,* 61–78. doi.org/10.1007/978-1-4615-0995-0_4

Sharpley, R., & Jepson, D. (2011). Rural tourism: A spiritual experience? *Annals of Tourism Research,* 38(1), 52–71. https://doi.org/10.1016/j.annals.2010.05.002

Sheldon, P. J. (2020). Designing tourism experiences for inner transformation. *Annals of Tourism Research,* 83, 102935. https://doi.org/10.1016/j.annals.2020.102935

Shin, W. S., Shin, C. S., & Yeoun, P. S. (2012). The influence of forest therapy camp on depression in alcoholics. *Environmental Health and Preventive Medicine,* 17(1), 73–76. https://doi.org/10.1007/s12199-011-0215-0

Stebbins, R. A. (1996). Cultural tourism as serious leisure. *Annals of Tourism Research,* 23(4), 948-950.

Stringer, L. A., & McAvoy, L. H. (2013). The need for something different: Spirituality and wilderness adventure. *Journal of Experiential Education,* 15(1), 13–20. https://doi.org/10.1177/105382599201500103

Tam, K. P. (2013). Concepts and measures related to connection to nature: Similarities and differences. *Journal of Environmental Psychology,* 34, 64–78. https://doi.org/10.1016/j.jenvp.2013.01.004

UNWTO. (2020). *Tourism and Culture.* Retrieved November 07, 2020, from https://www.unwto.org/tourism-and-culture

Valentine, P. (1992). Review: Nature-based tourism. In *Special Interest Tourism* (pp. 105–127). Belhaven Press, London. https://doi.org/10.1016/b978-0-08-044644-8.50010-2

Vespestad, M. K., & Lindberg, F. (2011). Understanding nature-based tourist experiences: An ontological analysis. *Current Issues in Tourism,* 14(6), 563–580. https://doi.org/10.1080/13683500.2010.513730

8 Professional Challenges while Guiding Tourists

Learning objectives

After studying this chapter, you should be able to understand:

- The demands of difficult tourists and how to handle them
- The difficult situations that the group may face and how to handle them
- The nuances of handling questions from guests
- The role of a tour guide in conducting the tours responsibly.
- How to lead a group on a multi-day tour
- The challenges of leading guests in a post COVID scenario

A visitor arrives for an entertaining experience, and delivering the same is the tour guide's responsibility. Often there are individuals, who for a variety of reasons, make the guide's job difficult. A guide must handle these guests professionally without letting the group's experience sour. Similarly, the touring party may encounter challenging situations when they look to the guide, as a leader, to navigate them through. Finally, it is a guide's job to satisfy the curiosity of guests who shoot all kinds of questions. Handling questions is a crucial tour guiding skill. In this chapter, we will discuss some of these challenges.

Visitors hire tour guides to assist them at the destination. They might require guides to help them make sense of the place or help them with logistical arrangements like getting quickly to the attraction, or helping them with tickets, or selecting what to engage with. It is in their best interest that their tour leaders and tour guides succeed. However, occasionally, awkward guests need to be handled prudently.

Most people in the tourism industry believe that post-COVID will be a different scenario. Visitor expectations will have gone a significant change. Destination leaderships also have come up with protocols. Tour operators must be ready to handle these changes through their tour guides.

Demanding tourists and handling them

Guests come to have fun, and they want the tour leader or the tour guide to help them have a good time. Most of the guests do not expect a super expert. However, some tourists behave obnoxiously. As a professional tour leader or guide, one needs to distinguish an obstinate tourist from a tourist with difficulty. Obstinate tourists are hard to serve because of their personality or behaviour. At times they may appear rude, unsympathetic or impatient. On the other hand, some tourists might be facing some difficulty and may have some genuine grievance.

Skill and professional competence are not about carrying out the routine job of a tour leader but handling the extraordinary situations that raise challenges. Handling a tourist who is obtrusive, obstructive, obstinate and obdurate is a challenge. Unfortunately, he is part of the party and has to be carried along all through, yet handled carefully, so that it does not lessen the fun that other members have come expecting. Let us consider some cases of demanding tourists:

■ The wanderer

The wanderer is a perpetual loiterer causing trouble when they do not report on time at a coach and other members of the party wait endlessly. This frequent desertion is a common problem. There are two possible reasons for defaulting.

The first reason is that perhaps the tourist did not understand the instructions given by the tour guide. So, these are not wanderers. A tour guide should give clear and precise instructions. Please ensure that such instructions are understood. Reconfirm. Do not give approximate time frames like, "we'll collect back here in one and a half hours". Instead, say, "We'll gather here by 4.00 by my watch. It is exactly 2.50 now. Please match your watches with mine." Tour guides must ensure that tourists understand the instructions. Use their language if you can, or ask someone who speaks their language to tell.

The others are compulsive defaulters who are often late. They may disturb the whole schedule and maybe the whole tour. Finally, there might be perpetual late arrivers who might not pay heed to your requests or express an insincere apology. They must be handled resolutely. Initially, the guide may reprimand them privately. On a subsequent occasion, they must try to create some peer pressure. They may take up the issue in the presence of the whole group and tell them how their behaviour is ruining others' fun. Enforce a five-minute rule: "that the party shall wait only for five minutes and will leave behind the latecomers." To penalise a perpetual latecomer, see if he/she can be left behind at the hotel at the beginning of the day. Please note that the guide may not enforce this on tour and leave missing guests at some far-off point where it would be difficult to get back on their own. Further, it is also a good habit to arrange a wake-up call for the guests. The guides can also tell them a time ten minutes before their departure.

Typically, the tour guide should not search for the wanderers. They should not also send or allow others to go and search unless they know exactly where to find the missing guest. The problem may get worse, with more guests missing. If you have waited sufficiently long and the person is still not traceable, inform the tour company. You should also leave information with responsible authorities at the site. Further, make sure that the other guests feel that you are doing something about the problem; otherwise, the habit of coming late becomes infectious, and the others guests also become late back.

■ The humiliator

Some guests continuously try to humiliate the tour guide by asking them questions only to contradict them subsequently. They try to 'test' the guide and show them up. This type of guest contradicts almost everything the tour guide says. For some reason, they feel the need to disprove the information the tour guide presents. The coping strategy includes being determined and polite at the same time. Do not let them take charge by distracting you by engaging in prolonged discussions. This engagement might ruin the mood of others in the group.

However, this does not mean that the guide needs to be stubborn. On the contrary, if not sure about something, the guide must be generous to accept: 'maybe you are right....' Remember, it is not an occasion to establish intellectual superiority. On the contrary, a tour guide's prime responsibility is to ensure that the tour and good mood are unabated.

■ The braggart

Some guests try to monopolise the tour. They behave as if they know everything, and want everyone in the group to acknowledge that they know as much or more than the guide on just about any subject. They refuse to accept advice or information from others and try to control the tour time rather than letting the tour guide do their job.

The tour guides must ensure that they control the group and situation professionally. Most guests want the guide to control, whereas the braggarts will try to snatch that away from them. Guides must not relent. However, they must not lose their cool, refuse to enter into an argument and rebuff firmly any attempts to overpower them. Should they feel that it is becoming difficult to handle such a person, they may seek support from the group. Group members, in all likelihood, may explicitly or otherwise convey their displeasure with the braggart – which is usually good enough.

The comedian

Another challenge may come from guests who might be called comedians. This is a personality type who tries to seek the group's attention by engaging in silly things. Though they may not be noxious, they must be kept under limits. They believe that they are doing an exceptional service by keeping everyone happy and entertained. While that is not a bad thing, the challenge may be that nothing is considered solemn, sanctified, deified or off limits to their jibes. While some hilarity is acceptable, it cannot be allowed to undermine of the situation. As the group leader, the guide must advise them to appreciate the gravity of the situation. They are generally not harmful but must be cut in before they tread too far. Sometimes the comedians behave so because they do not want to be on tour, and they hide their anxiety with these jibes.

The enraged

Sometimes the guides may come across angry guests, who may convey their anger in many ways – they may accuse, attack, criticise and judge the leader and others; or, may use foul language with name-calling, cursing, and offensive comments; or, they tend to raise their voice, shout, throw things and sometimes become physically abusive. They require attention. As a professional tour guide, you must swiftly evaluate the reasons for such behaviour. Sometimes the grievance could be justified and include abandonment and neglect, unresolved complaints against the sense of entitlement, unaddressed personal issues carried over, or victimisation, discrimination or ill-treatment by the group members.

If the grumble is justified, the leader may rectify it. They must display compassion and understanding; suggest that they are concerned with the guest's issues. Sometimes it is good to transform anger into a 'healthy anger' by offering challenges and motivating them to use the annoyance for 'you can do this!' Sometimes the cause of annoyance is group dynamics. The tour guide must intervene and try to resolve issues. One of the tools is reflective listening, which politely repeats back what the angry person has said. You may say, "What I hear you talking about is..." (Then repeat what they have said in their words, not yours). You can go on from there in some cases and empathise.

While there is no harm in apologising, make sure they feel heard. If the situation gets out of hand or the guest resorts to violence, seek professional help.

The dispirited

Sometimes, members of the tour party display their apathy by dull and sad looks. Though they do not cause direct harm, they stand off to the side, occasionally stray, visibly sigh a lot, talk on their cell phones, and act miserable as if they are unnecessarily being held hostage to the situation. It could be because of their personality traits or them being introverts. Nonetheless, in such cases, the tour guide must try to ignite interest by making attempts to include them wherever and whenever possible in the discussion. However, one should not spend too much time bringing prisoners (of the situation) into the fold so that the others feel ignored. Instead, ask them questions throughout the day that makes them feel seen, like "Did you have your meal on time?" Alternatively, "I hope you are liking the weather in the city." Such small questions can make them feel that they are being acknowledged.

■ The recluse

Sometimes the group size is small, and the guests are shy of responding to a tour guide's questions to prompt them to think. As a tour guide, one must try to elicit participation in the discussion while the recluse will not be willing to establish eye contact or answer their questions. They are indifferent to the tour guide's invitation to participate in dialogue: all they put together is a slurred word or two. It would help if the tour guide tried to take them on board by helping them with their small logistical things to win their confidence. Figure out what interests them. Try to include those subjects in the discussion. Try to converse with them in the language that they are most comfortable in.

■ The drunk

Once in a while, the tour leader will have a situation where one or more guests come for the tour in an inebriated state. If guests are intoxicated to the level that their behaviour is annoying, the tour leader must take a stand not to include them. Very politely, ask them to leave the tour group and take some rest. If they do not agree and start getting aggressive, they must inform the tour company and, should need be, inform the local authorities (at the site) and the police. Do not engage in arguments or brawl with such a member on your own. Ensure that the other group members are least affected and kept away. They are here to enjoy, and the last thing they want is to spoil their day.

■ The tagalongs

Tagalongs are individuals who knowingly or unknowingly get tagged to the group, especially during sightseeing when the tour guide is interpreting a place or thing. We met these in Chapter 7. A tour guide should try to disengage them politely, without disturbing the other guests.

In some cases, anti-social elements like pickpocketers may try to tag along. It is the responsibility of the tour guide to make out such attempts and keep the group away from such harm as discretely as possible.

Handling difficult tourists

Handling these is often situation-specific. A professional tour guide must analyse the situation and choose a way out so that the guests do not suffer and enjoy what is due to them to the fullest. Some tourists are not problematic, but have some fleeting difficulty. The following should be kept in mind:

- **Be impersonal**: never be personal. If the guest shares some ideas that are not correct, help him to the right ones. Do not point out mistakes of the individuals. Asian guests mainly try to save face. Shaming them could be disastrous. You may say, for example, "I expected, as most other tourists feel the same way. Nevertheless, the fact is..."
- **Remain calm**: There is no need to get animated about anything. The tourist is often from a different cultural context. Allow them a chance. Do not take it personally. Often, when the tourist has a complaint to make about arrangements by the tour company, guides take it personally and try to defend the arrangements. While they represent the tour company, they are not the company. A patient hearing is almost always the best solution. Assure them to convey their message to the tour company.
- **Empathise**: Empathise with the guests. They are in a peculiar situation. Feel for them and try to put yourself in the tourist's situation.
- **Assure them**: Do not shrug off the responsibility or pass the buck. The tour guide should stand up in control of the situation. They need not fool around with possible solutions to the tourists' problems. Instead, firmly suggest what will be done to attend to the issues. Go beyond the call of duty to help and comfort visitors. Exude confidence. Assure them.
- **Be proactive**: Display enthusiasm and fix problems. Do not avoid or delay a solution. Instead, take tourists in confidence and promptly and efficiently solve the top problem priority.

Ali refuses to drive any further!

The coach was not heading anywhere. Ali, the pilot, took the keys and got down from the bus. Raj, the tour leader, requested that he continue with the tour but in vain.

It was a group of tourists from the US who were on a tour of Uttrakhand, the hill state of northern India. The party started in Delhi, and they were to travel on a coach for four days. Raj was a seasoned tour leader and a destination expert for Uttrakhand. The company asked him to accompany this group of 17 young boys and girls. The group was determined to leave no chance to enjoy this exotic trip to India. It all started well on a high note. The group sang and pranked as they drove. The group found exciting things on their way and would get down to experience the same and take a few photographs.

They had already stopped four times, including a scheduled stop during the last five hours and were two hours behind schedule. Realising the delay, Raj asked Ali to drive straight to the next stop without halting at non-scheduled stopovers. Having been politely told they could not break in between as they needed to go as per the itinerary, the group insisted and started asking the pilot directly. Ali refused to entertain their request, and this annoyed the guests. A heated argument ensued, and Ali stopped the bus at a shoulder.

Raj tried to intervene to avoid the heated arguments between Ali and the group. He asked the group not to talk directly to the pilot but failed. The arguments turned into a brawl and became uglier when Ali responded by shouting at one of the tourists, and in turn, the tourist got so offended that he punched him hard on the face. Finally, Ali got down off the bus and refused to drive forward.

Raj was caught in an embarrassing situation. He apologised on behalf of the tourist. Ali insisted that the tourist who hit him should come and apologise. The tourist refused to oblige.

Handling emergencies

By its very nature emergency is an unforeseen situation that causes stress. Caught unaware, a tour party is left wondering what to do. The tour leader must spearhead planning and work to come out of the crisis, while ensuring that the guests do not get too anxious. In the case of an emergency, consider the following three-step approach:

1. Pre-empt

There is no better way to handle an emergency than to anticipate it and have a rescue plan in mind. See what all can go wrong and try to pre-empt. Most possible crises can be avoided by checking arrangements in advance and reminding guests well in time.

Guest, who shoplifted

8

Equinox, a Bengaluru-based MICE company, organised a conference for GVK Pharmaceuticals Ltd in Whitefield, Bengaluru.

After the conference, Roy, the local tour guide, took the attendees for a post-event tour of Whitefield. The attendees went to the Phoenix Market City Mall for shopping, leisure and recreational purpose. One of the attendees, named John, was window shopping. He went inside a watch showroom, where he deceitfully kept a watch inside his pocket, assuming no one was watching him. One of the showroom employees caught him and called the police, to his dismay.

One of the ways to not panic is to have an idea of the kinds of problems guides and their guests may face:

Personal issues of guests

- Illness
- Accident
- Theft/ pickpocket
- Lost documents
- Teasing or skirmishes
- Sexual harassment

Tour related issues

- Itinerary not completed
- Guides not professional
- Attraction not accessible
- Tips not included
- Attraction closed

Transport

- Breakdown
- Not turning up
- Unsafe driving
- Improper seats/windowpanes/ etc.
- Late arrivals
- Air conditioning is not working

Travel-related issues

- Lost tickets
- Missed train, bus or flight
- Lost luggage

Accommodation related issues

- Facilities not adequate
- Overbooking
- Not as promised
- Hygiene and cleanliness
- Voucher not accepted
- Poor service
- Lift unavailable
- Porterage charges

Food-related issues

- Hygiene
- It takes too much time
- Poor service
- Special food needs
- Food not to taste

Larger issues

- Floods, storms, etc.
- Infrastructure failure- power outage, disrupted roads, rail tracks, etc.
- Strikes, rallies, calls for protests, etc.
- Terrorism
- Crime- abduction, etc.

Many of these challenges can be handled if tour guides are proactive:

- Before the scheduled arrival of the guests, the tour leader should visit the booked hotels and restaurants to check if arrangements are correct. For example, they should ensure that the rooming list is ready, making the check-in smooth and quick. They should inform the restaurants of the party's arrival and its preferences. Informing the hotel while en route about the group's arrival allows them enough time to prepare the rooms.
- Guides must have the contact details of the bus drivers, taxis, autos, and any other mode of conveyance; talk to them, remind them of arriving on time, follow up. Similarly, the guide should arrive early and check the coach – its

condition (seats, windows, etc.), equipment (like microphone, LCDs, etc.). Ensure a kit is there in case of any emergency or breakdown of the vehicle. A spare tyre or two are available. This is especially important in long road journeys. Before the guests arrive to board the coach, they should be briefed about the itinerary and other requirements.

- Professional guides must have contact details of all important touchpoints – at attractions, museums, shops, and keep them informed. In a typical situation, guides may tell them that the group is running late. In some cases, the guides can manage a late entry at the attraction (entirely possible in many cultures and countries) – and it leaves an excellent impression, especially on western tourists (who are otherwise disappointed by the fact that they are late and so will miss the attraction!)
- Guides should always carry helpful phone numbers – police, ambulance, doctors, tour operator, hotels and restaurant managers, among others.

2. Sort out

Despite all precautions that tour guides take, they must be prepared to face challenging and demanding situations. If things go wrong, they must keep their nerve and appear in control all the time. A confident-looking tour leader keeps the group assured. At the first opportunity, they must brief the guests about the emergent situation and as it evolves. They should not hide anything, nor exaggerate the situation. They must assess the damage and likely loss and share information about it with guests. Tell them the truth but remain optimistic and keep stressing the positives.

A professional guide is genuinely concerned about the guests and their welfare all the time. Difficult situations are not the time to put the blame on anyone. If the situation appears to be beyond the guide's control, they must immediately inform the tour company and the concerned authorities, especially the police.

Attempt service recovery. Service recovery is a genuine attempt on the part of the service provider to reach out to a customer who has had a negative experience, even in the case of inadvertent emergencies. A crisis leads to a situation where the guest is deprived of what was planned. So, the guide should try to make up as far as possible. Do not count the cost. The welfare of guests is of utmost importance.

It is strongly recommended that tour guides neither advise nor administer medication. If required, refer to a doctor at the first instance. However, this should not prevent a tour guide from administering any first aid. In fact, in many cases, tour guides' training includes modules on first aid and CPR (CardioPulmonary Resuscitation).

The sick guest

As the group started with the day tour, Adam, one of the group members from Australia, got ill. The group was visiting Hyderabad and was scheduled to visit Ramouji Film City.

Since the previous night, he had been getting constant stomach cramps. Instead of taking the help of a doctor, he took his regular medicines. Early next morning, he vomited and got sicker. Shyam, the tour guide, came to know about his situation.

Shyam is in a dilemma whether to accompany the group on the day tour or first take Adam to a physician for a check-up.

3. Report

The final step in handling emergencies is to report. After the group has recovered, the tour guide must report the incident to the tour operator. It is advisable to write a short report about the incident with relevant details, including names, times, corrective action and outcomes.

It must be clearly understood that the tour operator has the primary responsibility for the group and so must be informed at the first instance. Further, they are better placed to send support and reinforcements. Tour guides must follow their instructions. On their part, the tour guides must never desert the group. They are the touchpoint, the local support, and the visitors' last hope in these situations. They must not run away. Never leave the injured unattended. Call for medical help where possible.

One of the crucial duties of the tour guide as a leader is to take care of the guests in cases of unforeseen situations. The clients need a tour company and a tour leader because they are wary of untoward situations in an unknown place.

Handling questions

Many tour guides believe that their primary job is to speak. However, they often wonder how much should they speak? At times, there have been complaints that the tour leader did not speak much and provided petite information, whereas , on other occasions, guests have snubbed or ignored the narrations. Throughout this book, we have tried to rebuff a common myth that the tour guides are paid to speak, so they must talk endlessly. It has been advised that the tour guides make a pitch for the monument, site, artefact, or whatever they intend to interpret in a bid to ignite curiosity. Therefore, they need not speak very long on each occasion. However, the success of a tour guide lies in the amount of interest he or she can generate, which in turn is assessed by the number of queries that are returned.

Questions are more than welcome. Tour guides need to appreciate the guest by publicly acknowledging their asking questions. However, the technique is that they must not answer only the questioner, which might leave other group members to wonder what the discussion is about. Instead, the discussion should be transferred to the entire group. Thus the guide must unequivocally thank the guest who has asked the question, turn to the group and repeat the question so that everyone in the group can hear it. Then they must answer to the entire group. However, do not make the entire guiding experience a question-answer round and limit the number of questions or reply later.

Note three things.

- If the group size is small – say, two to five people – who are already close enough to pay attention, the guide may not have to repeat the question.
- The guide may not have to repeat a question in response to their answer.
- Guides do not have to repeat the question if it does not pertain to the monument/ artefact being shown. For example, if a tourist has asked the toilet's direction, the guide does not have to repeat it to the group. However, whenever there is a question regarding logistical arrangements, repeat the question so there is no chance of misunderstanding.

Difficult questions

There could be challenging situations. There will be occasions when you do not know the answer. Sometimes, guests may ask embarrassing questions, politically incorrect or otherwise difficult to explain. The guide must not tell a lie or fabricate a story in such situations. They must be tactful in answering- this is the test of professionalism. Note some of the following:

Questions for which guides do not have the answers

There will always be questions to which guides do not have answers. Such situations need to be handled gracefully. Guides should not speak on subjects they are unaware of or do not have the necessary information. It is acceptable that a tour guide does not know everything. If they are unaware of the answer, they may say, "I have been in this profession for a long time, but never has this great question come up. I'll definitely get back to you." This is a tried and tested way to handle a question. A three-step action helps.

8

1. Note the question and ensure everyone sees them noting it down. Guests appreciate seeing the guide take their question so seriously.
2. Tell the asker when they will get back to them with the answer. Be honest.
3. Get back to them at the earliest or promised time. In one case, a guest was delighted to have received an email from the tour guide with whom she had been travelling in India the previous week. She had asked a question to

which the tour guide did not have an answer. So he researched and wrote back to her.

Do not use a shortcut of deception or flimsy explanation or description. This is an opportunity to learn, expand knowledge, and impress the guests.

Another technique often used by professional guides is to reflect the question to the group. "Does anyone in the group know the answer?" When a guide allows the group to help, they will appreciate being involved and sharing their knowledge. Do not underestimate their understanding of things.

Compliment the questioner with all sincerity, though such encouragement may motivate your guest sometimes to ask wayward questions. However, handle that by saying, "That's a great question. I've never thought about it that way. Does anyone here have an explanation for that?" It may be noted that the two abovementioned techniques have been combined.

Questions whose answers are controversial

These could be questions about sensitive issues like politics, religion, or controversial local issues. For example, guests may ask about a dogma or a myth associated with the destination or the community. Such questions should be handled tactfully. If possible, the guide should avoid talking and answering questions about sex, politics and religion, and should not express a personal opinion on these matters. They should not make up an answer that they think the questioner would like to hear.

A guide may choose to avoid answers by remarks like "We are still discussing this issue..." or say straightforwardly, "If you allow, let's not discuss political issues...". Likewise, the tours guide may choose not to share their opinions while asking the group to reflect: "what do you think…". They may take a neutral position by presenting the contrasting views: "there are two opinions. Some people think..., while others think...". Last, they may qualify the answer by saying "in my opinion...". It is best to avoid such questions and lightly mention that the guide focuses their time and energy on their work-life and less on such topics. Give less importance to such questions and move on.

Personal questions

Sometimes the tour guides are caught in a situation where a guest asks about personal issues. Of course, as a professional tour guide, one is not duty-bound to answer these. However, expert guides often use humour to bail themselves out of these situations. For example, if somebody asked: "how much salary do you get?" A guide should not answer accurately; instead, give it a humorous twist: (ironically smiling) "Ah! We are a poor lot…depend much on your tips!'. Alternatively, they may choose to generalise the answer: "tour guides here usually earn a hundred dollars a day." Others may pretend not to have heard the question or avoid the question by saying "never mind". Some tour guides respond through a disbelieving look (that such a question is asked!) followed by an exclamatory "pardon me" or "excuse me". Finally, the tour guide may prefer a straight rebut, "Sorry, that's too personal!" or "if you allow, I would not like to answer that..." Most of the questions can be avoided with just a smile on the face and a simple phrase "moving on to the next site…"

Pinching questions

Sometimes visitors ask questions that pinch because they are straight questions that show the destination or individual in a poor light. Such questions often annoy the tour guide or test his patience. So, the best answer to such irritants is to remain gracefully composed and acknowledge the questions with a smile. The key is spontaneity and creative humour. Perhaps this is the only situation where a guide may fabricate a befitting reply. The tour guide should not ignore the question. Instead, they should acknowledge it and try to present the positive side of the fact. Of course, it would be great if a logical explanation could be presented, because this was probably an innocent question and not the intention of the guests to undermine anyone. So, it might be the best time to educate them also on the positivities of the subject.

General tactics to answering questions

Having discussed some typical questioning situations, let us look at accepted tactics to handle difficult questions. With time guides can anticipate questions. First, they must have a list of FAQs. Prepare answers. If guides have contemplated the possible questions, it will be easier to answer them.

If a guest asks a question about a hypothetical situation, the guide must avoid answering it. Similarly, if they do not know the answer, they may choose not to comment on it. However, in other situations, they may respond to the question and connect it to some of the points they have made during the commentary, such as "and it is also important to point out…"

Sometimes a guest may come up with a pregnant pause and prod the guide to speak. They must remain cautious not to get pulled into talking simply to fill the air. Do not answer for the sake of it.

So, to sum up, remember always to welcome the question. Remain graceful and composed even if the question is difficult or disgusting, awful or agonising. Also, never look down on a question as foolish or frivolous. Often humour will help a guide bail out of a difficult situation. Also, negative questions can be an opportunity to make a positive point about the destination, community or self. Moreover, a tour guide is not obliged to answer all questions.

Leading tourists with responsibility

Another challenge is to lead responsible tours. Responsible tourism is holistic. It has an encouraging effect on the relationship between local communities, tourists, tourism companies and the administration. One aspect of responsible tourism is sustainability. It means having a little adverse impact on the environment and respecting and preserving local cultures. In Chapter 2, we discussed how visitors are getting concerned and want to consume holidays more responsibly. They want their tour guides to lead them into responsible holidays.

Protecting natural resources, flora and fauna, the built heritage and sharing the benefits of tourism with locals are the primary goals of responsible tourism. The involvement of local communities in tourism leads to their sustenance through employment, and their skill development. This involvement guarantees future jobs and, in turn, helps in conservation. Addressing the local community's needs is a prerequisite to identifying with the ecosystems and contributing to preserving the natural heritage. A tour guide should conduct tours and encourage the visitor to behave responsibly. The tour guide is one of the links that connects the visitor with the destination and other stakeholders, more importantly, the host community.

■ Contribute to the economic development of the community

Tourism activities will alienate the host community if the economic benefits are not shared. The host community often faces the brunt of tourism while they reap comparatively fewer benefits if tourism is not appropriately monitored (Chowdhary et al, n.d.; Kainthola et al., 2021b). A tour guide should do their best to contribute towards this end. While things with local character appear more attractive to visitors, provide guests with information about the scale of poverty in the community. Then, motivate them by explaining the need to spend in the local economy. For this, encourage them to visit local eating joints selling local food and beverages, include local craft fairs in the itinerary, invite guests to buy crafts that are made locally, to bargain responsibly and pay fair prices for goods and services purchased locally. Additionally, tour guides should encourage guests to donate to local cause movements. Showcase local community projects.

■ Responsibility towards the natural environment

Climate change is a big challenge. Tour guides should contribute to the conservation of the natural environment. This is especially necessary when leading in natural spaces. They must educate visitors about what they can do and what they should not. We have discussed some of these issues in the sectionon leading a nature walk. During the sightseeing tour, keep a watch on guests' behaviour and gently point out misbehaviour. Speaking casually, share with guests the list of the community resources that are over-exploited. Advise them to take care and behave when they encounter animals and birds. They may enjoy nature but must not chase or touch wild animals.

When outdoors, advise them to stay on the trails; trails are designed to lessen the erosion caused by the transit of visitors

■ Respect local culture

Each place has its traditions and customs, which is the tourism product. The host culture is the essential context for a visitor's experience, and as such, these cannot be allowed to deteriorate because of tourist activity. Tour guides act as a bridge between local people and travellers. One of their important roles is to facilitate access to the local community and help them assimilate with the local community. This allows the tourists an exposure that helps develop their personalities, ideologies, and experiences with the locals, where both parties gain tremendously (Kainthola et al., 2021).

The guests even want tour guides to educate them on how to dress and behave appropriately, especially in villages, religious sites, and cultural areas. In addition, they must help guests respect local culture, people, and traditions. Residents speak the vernacular and live with their age-old traditions. For visitors, it is essential to appreciate the difference and respect the local way of living. This is what they have come to experience.

Other actions include leading small groups on fragile trails and environs, using local transport as far as possible, advising tourists to seek permission to take photographs or to record vlogs. Similarly, they must advise guests to practice safe and responsible sex. Guests must not take part in any transaction of a sexual nature or that may involve illegal merchandise or drugs.

Tour leading

In many situations, the tour guide's job is enriched by adding responsibilities for logistical arrangements for the guests. Thus, from monument or city guiding, the responsibilities are expanded to take care of guests throughout the multi-day tours – from receiving them to seeing them off. The tour guide's role thus graduates to the role of a tour leader. This section discusses the tour leader's seamless delivery of the trip. The section is organised in the chronological sequence in which the tour is carried out. The roles and responsibilities of a tour leader at each stage of the tour are discussed. Typically, there are two situations. We also restrict our discussion to leading inbound and domestic tours.

Before the group arrives

The tour leader is usually responsible for the group from arrival at the airport, seaport or train/ bus station. A tour leader must swing into action before the group arrives. They must oversee the preparations for the group's comfortable stay, especially as it arrives.

General preparations

The first and most important thing is to ensure the hotel arrangements. First, they must ensure that they have the rooming list for the group. Next, they must ensure that the hotel is booked and rooms as per the booking. For example, some tourists might have opted for single rooms, some others might have requested rooms on twin sharing, and others for a double room. The tour leader should visit the hotels and inspect the rooms personally if possible. If the hotel package does not include drinking water in the rooms, ensure drinking water is available in each room.

In some cases, clients pay separately for this. They may also leave a copy of the itinerary, especially for the first few days, with the hotel to be shared with the guests. Some companies insist that such information must be provided on company stationery.

The other important detail is the passport and insurance details of the guests. Tour leaders must have these details and carry them throughout the tour. This is required only in the case of international guests. They should have flight details of the group, especially if they are arriving in with different flights.

The third necessary preparation is readying with the details of the trip. First, have a copy of the trip brochure to know what has been promised to the group. Next, keep ready a copy of trip notes and read through them. Finally, the tour leader must have the necessary travel vouchers with them.

Most companies provide tour leaders with a manual that includes all necessary documents like insurance details, release forms, indemnity forms, accident/ incident report forms, and other details.

Sometimes the tour operators have a welcome pack for the guests. This will have a welcome letter, information booklets, copies of itineraries, souvenirs, and other gifts. The tour leader must ensure that they have a sufficient number of welcome kits stocked. In addition, ensure that the first-aid kit is ready.

Tour leaders must prepare their own tickets and travel documents, their insurance, credit cards, and other things. They must pack their travel accessories. Finally, organise their thoughts and composure.

Last-minute checks

Before the group's arrival, the tour leader must check for the flight/ train/ coach/ cruise arrival status. Inform the hotel accordingly. Confirm that the transfer vehicle is ready and has reached the airport/ seaport/ train or bus station. Check if any land only customers would join the group directly in the hotel. Inform the hotel accordingly. Explain to the hotel how to deal with such clients and what support may be extended to them as per the package.

Many companies require that tour leaders wear the company's signature clothing – T-shirts, caps, jackets, or something else. In that case, tour leaders must have neat and clean company-branded accessories. For the guests, the tour leader is the company, representing it all the time. The tour leader must carry a placard with the company name to receive the guests.

Covid-19 has not disappeared, but we are only learning to live with it, though the government rules are constantly changing. These are unprecedented times, so a bit of oversight is possible, which guides must be wary of. Therefore, do not take any information and situation for granted and check for the new rules daily before stepping out with the group. If you are not feeling well, there is no harm in getting your test done and reporting to the agency. If you get infected, you must not be travelling with the group. Do not take the situation lightly.

■ Transfers

The pickup and drop-off service provided by a tour operator to customers travelling to and from the terminal (an airport or railway station or bus stand) is referred to as 'transfers'. Often tour operators expect tour leaders to engage with the guests from arrival until departure and help them with transfers.

Transfers are about helping guests with their physical transportation and possessions, getting them settled at the hotels, or carefully dropping them off at their departure terminal. However, it is much more than that. Depending on the situation, transfers include emotional-cultural transfers and physical transfers. While physical transfers refer to helping with physically transporting the individuals, emotional-cultural transfers relate to helping with adjust to an unknown place with survival tips and dealing with cultural shock. The tour leader, often the first person to meet the visitors at the destination, is naturally positioned to be the friend, philosopher and guide providing the much-required support to settle – physically and mentally. Given a good start, tour leaders earn the respect of the party, making handling the rest of the tour easier.

Preparing for the transfers

The tour leader should visit the terminal and conduct a reconnaissance to learn about the layout, facilities, procedures, and formalities there. The procedure and formalities at an airport are relatively comprehensive and rigorous – especially in the case of foreign tourist arrivals. For example, learn about immigration, customs, baggage reclaim procedures. At most airports, visitors, including tour leaders, are not allowed inside or beyond a point in the airport.

Tour leaders should make necessary arrangements and build rapport with porters so that their services may be used if required to transfer luggage to the vehicle/ coach. They must also coordinate with the vehicle pilot to decide where the vehicle will be parked, when it will be signalled to arrive, and where it waits to pick up the party.

Receiving guests

The first impression may not be the last, but it is lasting. It matters. Tour leaders should make the necessary preparations to offset uncertainties and surprises.

Preparation

Confirm the arrival time of the train, bus, flight or boat as the case may be. If a significant delay is expected, please inform the other parties – bus/coach operator, hotel, tour operator, and others. Being late is an unforgiveable sin, and the tour guide must always be punctual. They must reach the airport or seaport at least 30 minutes before the arrival time and at least 15 minutes in case of a train or bus station, and ensure logistics and a last-minute change of details, if any.

They must talk to the coach pilot/ driver and ensure that the coach or vehicles for transfers have arrived. Confirm that the bus is okay regarding the number of seats, quality of seats, windows panes, and other details. Ensure that the driver is carrying enough water bottles. You may also place the welcome kit on the bus.

Tour leaders must present themselves professionally, so they must be well-groomed and dressed nicely. Look smart and be appear confident. If required, wear the company's branded apparel. Often, flights get delayed, and you have to wait longer at the terminals, and sometimes the tour leaders may take a nap. So, as the time to meet guests closes, they must ready themselves, wash or wipe their face, and look fresh, even at midnight!

Organise porters and trolleys or at least keep them on alert. However, note that these services are available in many cases at an additional cost which guests may not opt for. In such cases, guides must seek their concurrence. Identify an appropriate point to receive the guests. Ensure that once the group gathers, it does not obstruct the movement of other passengers and people at work. In the case of a train, if they know the coach number, they may receive the guests at the train platform. In the case of airports, identify a point where all passengers will pass by to locate them quickly. The tour leader may carry the placard with the company name and maybe with the group leader's name.

A tour leader's checklist for transfers include:

- The bus/train/flight number, scheduled arrival time/terminal number
- If possible, some contact number from among the guests
- Phone number of the bus/coach captain
- Phone number of the concerned person at the tour operator's office
- They may require some cash to pay for parking, porters, and the like.

As the group arrives

As the group arrives, the tour leader must meet the party with positive energy and display enthusiasm, for example, help with guests with luggage. Acknowledge them with a welcome smile – everyone individually unless it is a big group. Identify each one and cross their names on your list. You must have a complete list of guests.

Empathise with the guests. They might have travelled long distances and might be exhausted. Tell them about the facilities at the airport, like washrooms, money changers, tea/coffee machines, telephone booths, convenience stores, and others. Provide adequate time to stretch and use these facilities, if need be.

Sometimes the tour leader may not be able to locate a guest. They may be missing. If there is a case of a 'no show' client, the tour leader must make all the efforts to ensure the guest is not travelling. They must check with other members of the group. Use the airport information system to locate the client. If required, check the related carrier whether the client got on the flight or not. Alternatively, contact the local agent or the tour operator to confirm the status of the missing guest. All efforts to locate the missing guests must be made before the group leaves the airport, seaport, or bus/train station. If required, make necessary follow-up arrangements.

Ensure that all passengers have reclaimed their baggage. Then, distribute the baggage tags and ask them to put them on. Help with porters and trolleys if need be.

Lost baggage

A typical case is of clients' baggage going missing. The first thing is to file a Property Irregularity Report (PIR) immediately. Some airlines will provide the client with an overnight kit or a prepaid debit card with some money loaded or cash, or an amount the guest can spend daily until the baggage arrives.

All assistance to clients with lost baggage should be offered at the tour leaders' level. Sometimes they may ask an assistant to help the client with airport procedures while you may take the group to the hotel if it is taking a long time. Empathise with the guests in such a situation but let them be aware that the company is responsible in no way. Moreover, if the baggage is not found, let them know their options and tell them about the insurance coverage policy. Ask them also to contact their insurance agent and ask for further procedures. Finally, let them know you will help them locate stores to buy new clothes once you have free time from the group.

Arrival briefing

The arrival briefing is the first and most essential as this connects the guide with the group. The briefing should be short and to the point. Provide them with the information so that they feel safe and relaxed. If a guest misses the briefing for some reason, tour leaders must ensure that they find an early opportunity to brief them individually. The arrival briefing can sometimes be broken into two parts: at the first meeting and then while on the coach for transfers.

Introduce

Seek the group's attention. Once the group is there, stand up and straight and speak clearly. Tour leaders must ensure that everyone is listening and understands what they say. Next, they should introduce themselves to the group along with their nickname or 'calling name' and unequivocally convey that they are the tour leader for the whole trip. Finally, they should connect with the group by sharing personal information like their association with the tour company or their authority/experience with the destination.

Tell the group about the next agenda. Tell them how far the hotel is and how much time it will take to reach there. Try to assure the group that they are in safe hands and enjoy the trip while the tour leader takes care of all the challenges.

Currency exchange

Often, the first information that the guests want is about money changing facilities. Inform them about the going rate and the facilities available. For example, most airports have money changing facilities, but it is often costlier to exchange money there. Advise accordingly, in that case. Advise the group on the best way of exchanging cash and using credit cards and ATMs, traveller's cheques, and other instruments. You may advise guests to change money later to avoid delays at the airport. Tell them when they can next change the currency. Also, the tour guide should inform the guests if their company does any foreign exchange, and they can avail themselves of the services if needed.

Getting connected

Usually, guests would like to connect with family and friends as they arrive and inform them about their wellbeing. They will require connectivity. Most international airports today provide free wifi to incoming guests. They may connect over VOIP through WhatsApp, Tango, Facebook, and other apps. The tour leader may help the guests with the protocol to connect to free wifi. Some international airports provide welcome SIMs to visitors. Help your guests register the new SIM. Alternatively, advise them on obtaining the local SIM and the best bargain.

Leading to the coach and boarding

Once everyone has collected, the party is ready to leave the terminal. Call the driver to bring the coach to the designated point as you lead the group there. Give clear instructions as to where the vehicle would wait for the party. Help guests load their baggage on the trolley/with the porters. Walk ahead of the group, holding placard/prop high to be visible to the group. Guests should follow the tour leader to the designated pickup point. Take care if a road is to be crossed. Remember, some guests might have a different orientation for traffic (side of the road that they must keep).

Once the group reaches the coach, ask them to ensure that their baggage is loaded into the coach luggage compartment.

Onboard the coach

Count the guests again to ensure that everyone has boarded the coach. Introduce the coach captain. As the coach moves, the tour leader may start talking (if the time of the day is appropriate), but be very brief. Tell the group how much time it will take to reach the hotel. If the situation permits, the remaining part of the arrival briefing can be continued on the coach during transfer to the start hotel.

Further, if the tour leader does not foresee an opportunity for a next briefing at the hotel (before they check into their respective rooms), it may be done on the coach. Brief the group for preparation, if any, for the upcoming trip. Give clear instructions about the time and place of meeting after they settle at the hotel and come out again for the tour. Sometimes, it may be necessary to have an additional trip briefing. Advise the group when and where this will be held.

Orientation about the accommodation

Give guests an idea about where the hotel is located. If possible, give them hotel cards with the address and directions to get back. Advise them on safety concerns – until what time is it safe to be out at night. Tour leaders may advise guests how the hotel is connected by local transport like the metro or the local mode of transport like tuk-tuk, and on costs, if any.

Arriving at the start hotel

Once the group arrives, ask them to sit comfortably in the hotel lobby. The tour leader must contact the front desk and introduce herself. Match the rooming list available with the hotel to the one with you. In case of a discrepancy, get the issues resolved. In case of any changes, inform the company/ ground handler at the first opportunity. If the discrepancy is enormous, the tour leader must immediately contact the ground handler's office.

The tour leader should take charge of the keys. Hand over the keys to the group only after they have briefed them. Otherwise, they may skip the briefing and rush to their rooms. A small briefing before the group gets to the front desk may be sought if the hotel reception does not allow you to take charge of the keys. You should brief them on the following:

- Inclusion of meals or water bottles. Let them know if water bottles are paid for.
- Special features of the hotel; orient the group about the place's layout, including restaurants/ coffee shops/ bars. Advise guests on the location of other facilities available in the hotel and the timings to use such facilities.
- Advise guests on the location of breakfast and the timings.
- Explain to the guests the minibar policy of the hotel. The minibar is often one issue of contention.
- Help the guests with wifi passwords, if any.
- Explain to guests the evacuation plans in case of emergencies.
- If required, advise on the use of elevators.
- Advise them on the intercom use because different hotels use different protocols.
- Tour leaders may have to advise them on using TV channels. In some cases, while some channels are complimentary, some others are available on payment. Inadvertently, guests subscribe to these channels and must pay on checking out.

Tour leaders must create a positive impression on the hotel staff with their professional conduct. If the staff believes that the tour leader is in command of the group, they will come forward to help him. They would appreciate it if the tour leader was helping them (the hotel staff), and they could trust them with that.

Board

Seek information on the guests' dietary requirements, especially if some are vegetarians. Note if some of the guests have any food-related allergies or specific dislikes. If the evening meal is not included in the package, advise them on options. The tour leader should consider organising a group meal and suggest a meeting point so that the entire group can go to the restaurant together. Tour leaders are expected to join the group for their first night's dinner. However, this may vary from company to company.

Tour leaders availability

Tour leaders must be available at all times, and the group should be communicated and assured of the same. In case of any need, they should feel free to contact the tour leader.

Briefing, handing over the keys and guests going to their rooms with their luggage should not take more than 10 to 20 minutes. The group should know that the tour leader will be in the lobby for the next 20-25 minutes. Should any of them have any issues, they can get back to the leader. Also, the tour leader should share with the group their room number and the intercom dialling number so that guests may contact them. Tour leaders' phone numbers may be included in the welcome pack given to them. Share other contact details, if any.

After that, the tour leader may lead the group as on the day trips. We have discussed guiding relevant techniques in previous chapters.

Leading in a post-Covid scenario

Covid is once in a century kind of a global crisis. It may take some time to recover from this pandemic, and Covid has ramifications that impact our way of conducting business, tourist behaviour and tourism supply, including guiding visitors. Tourism is one industry that was severely hurt by the pandemic, and tour guides were immediately and severely impacted. As tourism is selectively permitted again, guides have realised that they need to operate differently in a new ecosystem. The pandemic has impacted different places differently, and recovery was tailored to their specific requirements. First and foremost, the tour guides must be familiar with the local public health guidelines and must be prepared to adhere to the same. It is their familiar turf, and they must be fully aware of sites' specific Covid-19 requirements and comply with them.

Before the tour

Contactlessness is the order of the day (Kainthola et al, 2021a). Discussions with the tour company or the clients should preferably be online or over the phone, circumventing the need for any face-to-face contact. Likewise, bookings must be made online as far as possible to avoid physical contact with cash and tickets, and avoid queuing at booking counters. Help and encourage guests to engage in cashless transactions. If they cannot be avoided, the currency and hands should be properly sanitised.

If the situation merits prior contact with the visitors, the guide should share information regarding the public health protocols that tourists are expected to follow. In addition, they must advise about necessary precautions to be taken. Tour guides must contribute to the test and trace efforts of the local authorities.

They must obtain contact details for all guests and their representatives. Such records must also be maintained as mandated by the local authorities. Different authorities have specified different periods until when the records must be preserved. They must also support the public health authorities protocols such as temperature checks. They must insist that all guests wear masks correctly.

As the tour begins

At the start of the tour, guides should brief the group about the safety measures even if they was shared earlier. Such information would include instructions on handwashing, hygiene, and respiratory etiquettes. They must also advise the group to maintain social distance and avoid physical contact with surfaces. They must let the group know how to hygienically dispose of used tissues, masks and cigarette butts. In addition, the guides must be aware of the public toilets, handwashing facilities, points where hand sanitisers are available, and share this information at the beginning of the tour.

Further, they must advise guests to cover coughs and sneezes with a clean tissue or use their elbow. Spitting is prohibited in almost all public places. It is difficult to stay away from the crowd in public areas. However, everyone must try to maintain a distance of six feet.

Similarly, they should coordinate with coach companies, private vehicles, and drivers to ensure thoroughly sanitised vehicles and that the vehicle carries disinfecting wipes, hand sanitisers, and paper towels.

On the tour

Tour guides must lead by example and maintain government-recommended social distancing between themselves and guests at all times. Keeping the social distance translates into smaller groups. In some places, the public authorities have determined the allowed group size. Even otherwise, tour guides should insist on having a smaller group. They should manipulate their stops such that crowds are avoided and the group can maintain social distance. If more than one group is present, the tour leaders should coordinate not to crowd a location.

Tour guides are also familiar with the place and know which parts of the town are crowded at a particular time. Therefore, they must make necessary efforts to negotiate crowds and avoid them if appropriate. For example, they must avoid taking guests to crowded restaurants, museums, galleries, and other in-house events and activities.

Another consideration is avoiding physical contact, and so handshakes must be avoided. The Indian government advises the Indian greeting of 'namaste', which is increasingly adopted worldwide. Similarly, the tour guides must adopt rigorous hygiene protocols, like frequently and thoroughly washing hands and waiting for hands to dry. They may frequently use alcohol-based sanitisers to indicate that the group should do the same. In addition, they must carry personal hand sanitisers and use them when washing hands is not possible.

Tour guides should normally wear a mask or visor, all the time. Not on the chin, or do not ever take it off to speak. However, even if wearing masks and visors is not mandatory, the social distance may still be maintained. Guides increasingly use remote audio devices such as Whisper, Vox, and various apps to maintain a physical distance. The device can also facilitate guides to converse without taking off their masks or visors. Catching any sort of disease is not only dangerous to your health but is also a blow for you professionally. After such a long break where the income was negligible, it is high time tourism comes back more responsibly and with more mindfulness.

Social distancing must be maintained on shuttle rides, boat rides, jeep safaris and ropeways. The guests, pilot and co-pilot on the coach must wear masks and visors all the time. The pilot and co-pilot may also wear gloves while on duty. As the guests board the coach, they must be provided hand-sanitisers. Avoid using air conditioners on the vehicles. If necessary, the same should be done without recirculating the air.

Key points

Tour guiding is a profession with its own set of challenges. Guest expect to have an excellent time, and they want a tour guide to help them. While some customers may have genuine issues, others may be obstinate. However, a professional guide cannot allow the tour to be held hostage to these disgruntled guests. They must continually strive to mollify these guests also. There can be different demanding guests, and a professional guide must understand the underlying reasons to address them. Demanding tourists may be classified as wanderers, humiliators, braggarts, comedians, enraged, dispirited, recluse, drunk or just a tagalong. The chapter discusses how to address each of these cases.

A tour guide may face unforeseen situations that cause uncertainty leading to a crisis. In case of emergency, a three-step process is advised. The first step is to try to pre-empt the crisis. One way to do so is to remain aware of possible problems and have a plan in mind. The tour guide must understand that a crisis could be in the form of personal issues of the guests, some tour-related issues, transport and travel-related issues, accommodation of food-related issues or more significant issues beyond the control. The second step is to sort out the issues. A tour guide must never abandon the group. They may attempt recovery but must not administer medication on their own. The third step is to report any untoward situation that might have arisen.

Another critical skill for tour guides is to handle questions. Guest can pose any questions for which the tour guide does not have answers, those that deal with controversies, personal questions, or pinching questions.

One of the essential duties of tour guides is to conduct themselves and tour groups so that tourism has minimal negative impacts on the host community and their environment. Such tourism is responsible tourism. Broadly these include economic, cultural and concern towards nature.

We discuss the role of a tour guide as a tour leader for multi-day tours. The role enlarges to include logistical arrangements for guests in addition to interpretation. Tour leaders must prepare to receive the guests, transfer them to the hotel, and settle them in their allotted rooms. Helping guests with lost luggage, helping guests board on the coach and briefing them at opportune times. Tour leaders must ensure that they are available to the guest all the time.

Finally, this chapter discusses leading a group in a post-Covid scenario. Covid has posed severe challenges for tour guides. It will change tour guiding for all times to come. Health, hygiene and safety protocols must be built into every stage of tour guiding – before and on the tour.

Key terms

Braggart	Try to monopolise the tour. They act as if they know everything and want everyone in the group to acknowledge that they know as much or more than the guide on just about any subject.
Comedian	Seek the group's attention by engaging in silly things, apparently believing that they are doing an service by keeping everyone happy and entertained.
CPR (Cardiopulmonary resuscitation)	A lifesaving procedure useful in many emergencies, such as a heart attack or near drowning, in which someone's breathing or heartbeat has stopped.
Dispirited guest	Apathetic guests who stand off to the side, occasionally stray away, visibly sigh a lot, talk on their cell phones, and act miserable.
Enraged guest	Guests who are upset and who may convey their anger in many ways towards the leader and others; they may use foul language, shout, throw things and sometimes become physically abusive.
Humiliator	This type continuously tries to humiliate tour guides by asking them questions only to contradict them subsequently. They try to 'test' the guides and show them up.
Recluse	Guests who are indifferent to the tour guide's invitations to participate in dialogue: all they put together is a slurred word or two.
Tagalong	These individuals knowingly or unknowingly get tagged to the group, especially during sightseeing when the tour guide interprets a place or thing.
Wanderer	A perpetual latecomer causing trouble when they do not report on time, so that other members of the party wait endlessly

Discussion questions

1 Talk to some local tour guides – note challenges they face while dealing with different types of guests.

2 How would you deal with a tagalong who pretends not to understand that yours is a private group?

3 A guest is repeatedly asking about the poverty in the slums of your town to show the place in poor light. How would you handle such a situation?

4 Instructions from the tour operator said that the train would leave from platform no.5. However, as you arrive with guests, you learn that the train will leave platform no. 15. What will you do?

5 As you arrive at the airport to drop the group off, you realise that the flight is delayed by 7 hours. What will you do?

6 The coach that has come to pick up your group at the airport has broken panes. What will you do?

7 Mr Jones from South Africa, a member of your group, informs you that he has lost his passport. What will you do?

8 On arrival at the hotel, your group complains that the hotel rooms are not their liking. What will you do?

Activity

1 Form a group of four people: three guests and one tour guide. Create an impromptu situation for the guide where they have to take care of the grievances of the guests.

2 Solve the case studies provided in the chapter in your way.

References

Chowdhary, N., Tiwari, P., & Kainthola, S. (n.d.). *PRME: The Way Forward to Deal with Overtourism and Related Perverse Impacts.*

Kainthola, S., Tiwari, P., & Chowdhary, N. R. (2021a). Overtourism to zero tourism: changing tourists' perception of crowding post Covid-19. *Journal of Spatial and Organizational Dynamics*, 9(2), 115–137.

Kainthola, S., Tiwari, P., & Chowdhary, N. R. (2021b). Tourist guides' perspectives of demarketing the Taj Mahal. *International Journal of Tour Guiding Research Volume*, 2(1). https://doi.org/https://arrow.tudublin.ie/ijtgr/vol2/iss1/4 This

Kainthola, S., Tiwari, P., & Chowdhary, N. R. (2021). Deconstructing volunteer tourism. In R. S. K. Holmes, L. Lockstone-Binney, K. A. Smith (Ed.), *The Routledge Handbook of Volunteering in Events, Sport and Tourism*. Routledge.

9 The Business of Tour Guiding

Learning objectives

After studying this chapter, you should be able to understand:

- How to set up a tour guiding business
- How to plan an itinerary
- The importance of various partners in tour guiding business
- The spirit of relevant laws and the legal considerations in tour guiding
- The importance of ethical conduct

Most tour guides are freelancing professionals who associate themselves with tour operators. A guide may associate with more than one company to broaden their market, earn more income, and reduce risk at any point in time. However, there are also companies that offer the services of professional tour guides. They deploy different business models to position themselves. Some companies enrol and promote tour guides that are more comfortable with digital connect. They profile the registered professionals on parameters, rank them, and sometimes offer some orientation or training before connecting them with tour operators (business-to-business) or presenting them directly to the tourists (business-to-consumer). They provide information on the tour guide's USP (unique selling point), language proficiency, and other relevant skills.

Some tour guide companies meticulously curate experiences and offer the same through professionally trained tour guides. These companies invest in research and curating theme-based experiences, but also put considerable effort into recruiting and training their tour guides. A variant that has become popular these days are the companies that organise walking tours. They often create a limited sightseeing event based on local culture, heritage, or simple contemporary life. For example, there is a husband-wife team in Morroco which offers the popular food trail called Marrakech Food Tours. The visitors get the opportunity to taste slow-cooked lamb, traditional salads, pastries and the famous mint tea. India City Walks is a much-acclaimed company that curates a range of experiences and operates in several cities. They also invest time and effort in training walk leaders.

Guiding companies are popular in niche operations like nature walks, adventure trails, forest bathing, and wellness therapies. Especially in areas where the tourist finds it simpler to go via an established channel or is unaware of that destination. However, technology has empowered individual guides, and many of them operate as one-person companies. This chapter is about the nuances of tour guiding businesses – setting up, designing an itinerary as an essential product, networking, underlying ethics and principles of conduct and other issues.

Tourism has witnessed a paradigm shift. Visitors today are less interested in visiting the identified popular tourist spots at a destination. Instead, they seek immersive, unusual experiences. Any destination has many resources. Locals are advantageously situated to curate experiences based on tangible and intangible resources at hand. It could be a simple culinary experience or a walk through a haunted quarter. Demand for customised and unique experiences is on the rise. It may be an excellent opportunity to be self-employed and monetise one's passion for showing around their place with such a demand.

Communities and governments have also realised the importance of entrepreneurship. They support startups through policy and institutional interventions. For example, India's government brought in legislation permitting the creation of a company with just one director and one member. In these companies, compliance requirements are lesser than that of a private company.

Setting up a business

It is a good idea to start a tour guiding business in a popular tourist destination or one which has potential in the near future. However, it is better if the guide is familiar with the place – its history and geography. Thus it is always advantageous to have a guiding company in your native place where you can have a good relationship with the local community. A personal tour guiding business can be started with less than USD 2,500 and earn more than USD 100,000 a year. Tour guiding is a business based more on skills than on investment. There are multiple revenue streams for a tour guide, and there are many examples of tour guides doing exceptionally well financially. A good marketing plan and networking is the key.

Some key steps in setting up a tour guiding business include the following.

Check the requirements

In most places – especially established tourism destinations – the authorities have a protocol for starting businesses, including tourism-related ones. Therefore, the first step is to check the essential requirements for setting up a tour guiding business with the local authorities. Some documentation will be required for registering as a business entity. In many places, individuals operating as tour guides must obtain a guiding license. For instance, a tour guide must have a license to guide in sites under the Archeological Survey of India in India. Find out if it is required in your case. Contact the concerned authorities in your vicinity and learn about the process for obtaining a license to be a tour guide.

Market research

It is essential to understand the market that you want to be in. Research the competitors and the customers. Find out what kinds of tours are being sold in the area. Try to understand what is being offered – and what is not. Is there a new emerging market, or are the visitors increasingly demanding particular experiences. It may be good to take a few tours in the area to know what kind are offered. Either you may decide to offer the same tours to the same clientele, or you may have a creative idea to offer a differently themed tour to the existing clientele or a speciality tour for a niche market. What can be presented to tourists: monuments, religious places, the life at the place, art and shopping or cuisine, history or nature? Find out what differentiates you from the existing groups and how you can sell your product in the competitive market. If your idea is not already present in the market, then analyse why that is the case. Is it because it is new or because someone has already tried and it did not work out. Find out what works and what does not. Make a long term plan.

One of the keys to success is to develop a USP that can be defended over a period. It essentially reminds you why the patrons should choose your tours over those of the competitors. As long as this difference is evident in the customers' minds, they will continue to buy your tours.

As long as your customers believe that yours is the only tour of its kind in the city, or that you offer more value at comparable or lower price, or you tell engaging stories or present a different perspective or a different aspect of the community, they will positively differentiate your offer. Most of the products offered depend on the locally available tangible and intangible resources – natural, built, or cultural environment. Therefore, it is advised to carry out a thorough recce of the place. First, decide about the product portfolio. After that, conduct further research, make notes, and identify sales pitches – take note of odd, unusual, or fascinating information.

Select the target market

Based on your research and ensuing deliberations, decide the market your tour guiding company would like to serve. Generally, there is more than one type of visitor to any destination with different expectations. Along with the primary motivation to visit the place, they may seek various experiences: eco-trail, culinary walks, photo-walks, and heritage walks that focus on some lesser-known aspects of the host community. Anything could potentially be a tour.

Understand which audience can be attracted to your product and focus your marketing and selling on them. You can either make your product a niche or incorporate different sections of society. However, it is essential to have an organised plan for marketing. For instance, photography tours are more for the photography enthusiasts, or hiking trails are more likely to be attended by the young crowd. Therefore, focusing the efforts on them is important. You may choose to provide guides to a tour company, or deal directly with the clients, or do a mix of both. Marketing and selling in both cases are very different.

Form a company and brand

One of the first decisions is to have a name for the company and the tour products. The name is often an irreversible decision that has a far-reaching impact on the business. It must appeal to and attract customers who relate it with your offers. It must be simple yet unique and attractive. Often, the clients are from a different cultural context, so the name must be comprehendible. They must effortlessly pronounce, recall and write it. Be creative in choosing a name.

The next step is to register the business. You should approach the relevant local authority to register the company legally. At this stage, you should also decide about the form of organisation. You may prefer forming a Limited Liability Company (LLC) to keep the business separate from personal liabilities and spread the associated risk. In India, a variant of LLC is called the One Person Company (OPC), which is functionally a company with only one shareholder as its member. As discussed earlier, obtain necessary guiding licenses wherever required. You should complete this process, and while waiting for the necessary permissions and registration documents, plan for other things.

Once the company is registered, a separate bank account should be opened. This account must be used exclusively for business transactions. In addition, this account must be separate from your personal account.

Record of start-up costs

Work out your startup costs and keep a record of the same. Initially, be conservative and do not spend much money on things not required immediately. Initially, most costs are associated with startup formality costs, setting up the office, and initial marketing expenditure.

Some initial costs include the cost of registration and paperwork. Fees of consultants who have helped you with registration. List props and tools (in the kit) that you will require during guiding or back-office like computers, books, videos, and others. You may also like to have a web presence. You may have a website, a registered domain name and an associated email which are part of the branding efforts. The entrepreneur must get an attractive logo designed. Have visiting cards, brochures, and pamphlets highlighting your tours' theme and unique features. Get expert help in designing. The entrepreneur may factor in these costs.

Begin marketing your tours

Marketing is crucial to the success of the business. Tourism is essentially a networked business, so partnering and selling through the network is critical. First, identify the important tour companies, including ground handlers who operate local tours on behalf of wholesalers. They are the principal clients who hire your services for their guests/ parties. Discuss your tours and understand their tour ideas. Then, negotiate terms and conditions and fees for your services.

Talk to tourism authorities in your city and share your plans to start tour guide operations. Inform the local police about your intentions to start the operations. Restaurants, hotels and other utility providers along your routes are important network partners. Connect with them to cross-sell. Ask them to provide you with their brochures so that you can share information with patrons. Similarly, touch base with local transporters – taxis, autos, coaches and others.

Create an online presence. Create a website and pages on social media platforms like Facebook, Instagram, and Linkedin, among others. Next, generate content to engage with your potential audiences. Understand your audience and then choose the right way of marketing. See what attracts them and which mode of communication best suits them. For instance, elderly clientele connect more through the internet and direct marketing. In comparison, the young crowd is attracted more by social media and reviews. Increasingly, companies are using cross-platform distribution, posting content across multiple platforms, including their website, blog, and various social channels. It is an excellent strategy to engage with travel influencers on social media. They may help you promote your tour guide business. Sometimes, companies may like to use an online booking tool to make it easy for customers.

Keep it simple

A popular theory of entrepreneurial startup is the theory of effectuation (Sarasvathy, 2001). One of the principles propounded by her Bird-in-Hand suggests that to begin with, entrepreneurs need to create solutions with the resources available here and now. A man can begin minimally as just him. However, they may need a computer, phone, and internet connection for booking. List the props and tools (in the kit) that you will require during guiding. Procure the books and videos that you must possess to prepare for commentaries.

The company may decide to have offers that may require some assets. For example, if the tours are designed to be conducted on bicycles or Segway, equipment will have to be procured. In addition, the tour company may decide to expand their services to transfers or arrange for a vehicle for sightseeing. These costs must be factored into the financial planning.

It is good to take some lessons on keeping personal and business accounts.

Write a business plan

A business plan is a written document that roadmaps how a startup will achieve its goals. This serves two purposes – it helps an entrepreneur conceptualise the business; and the progress can be guided vis-à-vis the plan later.

There are several benefits of writing a tour guide business plan before actually starting with the business:

- Have an estimate of your proposed business and if it will be feasible.
- Bring together all your ideas of products with their unique features in an organised format
- Put your vision for the business to share with others within the company and outsiders who will matter: travel company, other trade partners and lenders if required.
- A business plan estimates and forecasts future possible scenarios and challenges, and tries to address them before they threaten your tour guiding business's success. In addition, it may outline alternative plans.
- Set out the business strategy: who will the customers be, what will you offer them, who are the principal partners, how will you relate to them?
- Set targets that will serve as benchmarks to help you monitor the business performance continuously.

A good business plan includes an executive summary, company description, market analysis, products and services, marketing strategy and analysis, human resource and operations plans, financial planning, and a budget.

Components of a business plan for a tour guide company

Executive summary: A summary of all the elements in your plan, tours and their concept; tour guiding operations; marketing; the staff you might employ; the rationale of all these and costs and therefore the finance that you would require – your own money and the money that you will have to borrow.

Market analysis: Research the potential market for your services, the visitors, how they arrive at the destination, whom (agency), and how much you will have to spend to get access to these guests.

Competitor analysis: What is the competition? Fellow tour guides or technology in audio guides/ guide books, among others. What are their strengths? Can you match these? What are their weaknesses? Can you come in? What is that you can do that the competitors cannot do? Who are the main competitors – other tour guides and the tour companies.

Sales forecast: Try to work out your sales over the next year, or maybe two to three years.

Company management: What will be the legal status of the company? What will be the organisation of the company?

Operations: What will be the details of the tour? The duration, length, and resources that you would require. Identify the costs associated with these resources.

Staffing: Will it be a one-person company or have other colleagues and support staff? Work out your potential staff's job description and the number of people you need to work for you. What are your plans to add staff as the business grows?

Financial statements: These include a balance sheet, operating expenses, other expenses, and business income. It is best to consult a CA or an expert. Yours is a specific business, and your businesses accounts will be straightforward to keep. Specific information to be used to figure out these costs will include your estimates of:

Revenues	The average number of groups handled per day × number of working days in month × fees collected from each group + income from other business-related sources
Variable costs	Average cost per unit (per group or head) × Average number of groups handled per day × number of working days in a month
Fixed costs	The cost of premises, equipment, permanent staff and other expenses that arealways there
Gross profit	Revenue minus the variable costs
Net profit	Gross profit minus fixed costs
Net income	Net profit minus taxes

Planning an itinerary

A tour guide's offer may be conceptualised and presented as an itinerary – a plan of activities related to a trip. Often, the tour operator designs a comprehensive itinerary, and requires tour guides to execute parts of this at selected locations. In other situations, tour operators seek creative sub-itineraries promising unique experiences from ground handlers or directly from tour guides or tour guiding companies. Within their larger itineraries, they accommodate individual providers' itineraries. Further, a tour guide may suggest changes to the tour operator in the itinerary if a different route is more feasible or exciting.

Usually, the travel agent has a set of well researched, pre-planned travel itineraries to offer. Visitors may pick the one they find closest to their requirements, and sometimes there is a choice to customise the itinerary. However, with easy access to online maps, navigations, aggregators, and peer-to-peer trip advice, the younger generation is inclined to design their own itineraries. Still, many people approach tour guides for travel assistance and help execute itineraries. Guides, the local experts, assist in making the trip efficient and effective. They are efficient because they know the place and protocols and are better placed to help avoid time wasters. They can advise the shortest route, best times, and negotiate with the locals and authorities at particular attractions, and at the destination in general. They make the trip effective by providing timely and accurate interpretation, which the visitor might miss if on their own.

The itinerary includes spots to be visited at specified times and sometimes transportation between those attractions. Travel itineraries have three dimensions routing, scheduling and loading. They also spell out the modes of transfer.

- *Routing* is the process of creating the most cost-effective route through the minimisation of distance or time necessary to reach a set of stopovers. Routing prescribes the path and sequence of places to be visited at a destination or an attraction. It is a necessary process, primarily due to the high competition and narrowing margins in the global tourism market. It implies monetary and non-monetary costs for the patron. Monetary costs may include vehicle operation, fuel, labour, and maintenance. Non-monetary costs for the visitors may include physical effort, search at a destination, opportunity loss due to time spent waiting, and missing out on essential attractions or spots.
- *Scheduling* assigns arrival and stayover time at each stop – when will the guests arrive at a place, and how long will they stay there.
- *Loading* spells out what activities are to be carried out at each location. It determines how much time are they going to spend at these. The activities may have to be negotiated beforehand and even staged.

Principles of itinerary design

Some of the principles to be kept in mind while designing an itinerary are:

Principle of downtime

For an efficient tour, the downtime (or transit time) vis-à-vis the activity time should be minimal. Travelling is physically demanding, so one should have the target audience in mind. Elderly guests prefer a slow tour (transit time allows them to rest and recoup). A younger group might rush through and make most of the tour by visiting as many places as possible. In some cases, destinations impose restrictions. In Israel, for example, transit time is restricted to 10%, a maximum of 200 km and 8 hours of touring time per touring day.

Downtimes are essential because they allow for social bonding and relaxation. Many memorable experiences are formed during these 'unscripted' parts of the tour. However, they must be restricted else the guest is dissatisfied, having spent a lot of the time in transit.

Principle of flexibility

The designer may like to include some time flexibility on the itinerary. Some guests may feel too restricted by the time constraint, while others demand this structure as a part of their package. All guests will require some time to feel the place and internalise the experience their way. Therefore, building extra time in an itinerary is usually a good idea. Then there will be some unforeseen loss of time due to incidents, some confusion, or some accidents. A time cushion will accommodate all these losses and still complete the scheduled activities.

Principle of pleasure

A tourist travels to enjoy, so the trip must be entertaining, with the minimum of hiccups, bottlenecks, and restrictions. The travel must be convenient and comfortable. Tight schedules and physically demanding tours may be stressful for the guest. Too many dos and don'ts cause anxiety. They must be avoided.

Formatting the itinerary

Presenting an itinerary could be pretty challenging. Therefore, it is suggested to use a visually appealing and organised format with direct and less fancy words.

- Put a title to the itinerary with a small, nice-looking name at the top with the destination and travel dates below the title.
- Make a sub-head titled 'Trip details'—split details into sub-categories arranged by date. For example, divide the itinerary by day or time of day.
- Create a column for each sub-category with hours on the left-hand side.
- Divide the itinerary by day, then break the day into parts. Pin each part to a stop and allot time to each. Finally, organise icons and accompanying information and time within each sub-category on the itinerary.

- Use start times (or approximate times) to organise each part of the day.
- Having a quick run of the day's events will help the company. Use a format similar to the one below:

7.00 am-	Walk to the local restaurant
7.30 am -	Breakfast in a local restaurant offering special local tea and snacks (Jalebi[1] and Kachori)
8.30 a.m. -	Departure from restaurant (by bus)
9.30.00 a.m. - 11.30 a.m.	Visit Amber fort
11.30 a.m. – 12.00-noon	Travel to Jantar Mantar
12.00 noon – 1.00 p.m.	Visit Jantar Mantar
1.00 p.m. -	Lunch at *Copper Chimney Restaurant*

- Space out events on the page for easy reading. Create a tabular structure so that the reader's eyes will go to the most logical places (dates, times, vital info in that sequence). An excellent way to do this is to left-align the start times for activities and use tabs to space out the text.
- Include images to show what type of travel information is presented for that day. For example, use an image of a car to indicate a taxi or rental car, an image of the house for a hotel, and similar signage. Use popular and catchy images that the visitors can readily recall and relate to. Use pictures that develop the urge to travel in them.
- The company may prepare two versions of the itineraries: a condensed version for the guests and a full version for internal use. The guest versions can show times and places or points of interest. The internal use versions will have as much information as possible: the notes about the attraction, phone numbers, addresses, email addresses, bus or walking, anything that the tour guide can think of.

Other issues related to itinerary design and development

1 Prudently choose a title/name for the itinerary. Besides being nice-looking, it should precisely reflect the character of the proposed itinerary.

2 Price the itinerary. Take into consideration the per diem rates, if any, fixed by the relevant authorities.

3 Review the itinerary a day or two before the trip. If the company has planned visits to a monument or a resort, call or email to receive confirmations. Talk to the tour operator; the travel plan often changes, and nobody wants to miss the boat. Make sure that all the activities in the itinerary can be done. If there are any chances that some activities can not be done, add a disclaimer – subject to weather conditions or availability of tickets.

1 *Indian sweet pretzels (for lack of a better description)*

Example of an Itinerary

THE AWESOME AGRA

Destination Covered: Delhi - Agra – Delhi

Tour detail: (A day trip to Agra - Daily except for Friday, Taj Mahal is closed on Friday)

Agra is a major tourist destination because of the Taj Mahal, one of the seven wonders of the world and the magnificent Mughal infrastructure, arts and culture. Sikandar Lodi of the Lodi Dynasty established Agra city in the 16th century, and later it became the capital city of the Mughals. The beautiful Taj Mahal in pure white marble took 22 years and 22,000 artisans to build. Emperor Shah Jahan built this monument in memory of his beloved consort Mumtaz Mahal. We will also visit Agra Fort, whose maze of the courtyard, mosque, and private chambers reflect the Mughal Empire's grandeur.

The private guide will be accompanied by a chauffeur-driven AC car to make this trip comfortable and enjoyable at your own pace with time for photo stops and shopping.

Attractions – Taj Mahal, Agra fort and Local Art & Craft, and many more

06:00 Hrs Outbound	(Delhi to Agra by car/230Km approx. 3.30hrs) You will be picked up from the hotel and driven to Agra through Yamuna Expressway. Half an hour for breakfast at mid-way restaurant.
10:00 Hrs Taj Mahal	Reach Agra. You will be taken to the most famous attraction of Agra – The Taj Mahal. You will get the time to explore its stunning beauty.
Noon Agra Fort	After visiting the Taj Mahal, you will visit the Agra Fort, the magnificent beauty of red sandstone positioned on the Yamuna river bank.
13:30 Hrs Lunch	After visiting Agra Fort, enjoy a delicious lunch in a local recommended restaurant or hotel.
15:00 Hrs Arts & Crafts	After lunch, we will visit some very famous local art and craft. Agra is famous for inlaid marble work, carpet weaving and Zardozi work.
17:00 Hrs Return	In the evening, proceed to Delhi through Yamuna Express highway and transfer to the hotel with sweet memories of the same day Agra tour.
20:00 Hrs	Once you reach Delhi, you will be transferred to the hotel.

Price USD 110 Per Person (for a minimum of 02 travellers)

The price includes :

All transfers and sightseeing as per mentioned itinerary by AC car,
Local English speaking guide,
Meals – Breakfast & Lunch (in local restaurant)
All currently applicable taxes.

The cost does not include:

Entrance fee to monuments,
Camera/video fee, drinks, tips, any other item not mentioned in supplement for entrance.
Any personal expenses

Pricing a Package

Price is the cost to the customer. From a supplier's point of view:

Price = Costs of goods and services + profit

Sometimes costing is simpler as the itinerary is short and straightforward, say a one hour tour of an archaeological site. In other cases, it could be more extended tours over multiple days, and with other elements to be accounted for.

Determine costs

One can look at costs from two perspectives – the nature of costs and elements of costs. There are two types of costs: *fixed* and *variable*, though some fixed costs might vary. These may be called *variable-fixed* costs.

Fixed costs

These are expenses that will remain constant throughout the trip, whether there are 4 or 6 or 10 people in the group. The tour manager and guide's expense will not change with the group's size (up to a limit). Fixed costs also include *indirect* costs, which are expenses not specifically for one particular tour. For example, a package's promotional cost would be an indirect cost because the company will use it for more than one trip. Overhead expenses like office rentals, interest on investment in equipment, consumables, etc. will not vary. The company must determine a formula to apportion part of these expenses to the tour costs.

Variable costs or direct costs

These are expenses that vary on a per-person basis. Every guest means some additional cost. Thus, admission fees, food expenses, and memorabilia costs are variable costs. The total amount spent on each item will vary according to the number of people participating in the tour. When compiling the trip's cost, you will need to calculate a total for all the variable costs associated with the tour.

However, the company may negotiate special rates for these expenses. Partnerships between companies may help in this situation. Suppliers may provide special discounts to be on the guide's package.

Variable-fixed costs

Some expenses do not vary per person. However, some vary as per the policies of the provider. An example of variable-fixed costs is local transportation. Often the coach will have a fixed capacity. In India, there are small coaches with 12 or 18 person seating. Beyond this, a larger vehicle will have to be engaged, accommodating 28 or 32 or 40 Passengers. If the price for a 12-seater vehicle is INR 16 per km, the cost is fixed irrespective of group size up to 12 people – whether it is 6 or 8 or 10 guests. An 18-seater coach may charge INR 18 per km. So, if the company has 13 guests, the fixed cost will increase immediately. The transport cost also varies with the distance (per km). Since the itinerary is fixed, the transport cost can be considered fixed. For example, if the point-to-point distance during a sightseeing trip is estimated at 65 km, it may be increased to 75 km to take care of any deviations. So, for a 12-seater, the fixed costs will be INR 1,200 (75 km × 16 INR), while for an 18-seater, costs will be INR 1,350 (75 km × 18 INR).

When determining costs for the tour package, the company should not calculate a total for the changing fixed costs because each item will change with the number of guests in the group. If the company's policy restricts group size to less than 18, it may take the city-tour transport cost as INR 1,350.

Elements of costs

We have discussed the components of an itinerary. Most of them will have a cost implication. Let us discuss them one by one.

A group of 15 students from a college in Delhi plan to visit Pink City Jaipur for a day. They will reach Jaipur the morning of a Saturday and return late evening. They have approached your company "The Jaipur Guides (TJG)" to organise this day trip.

Meals

Some restaurants cater to visitors. They also negotiate special rates with travel agents for groups.

In this case, students will be visiting the local attractions, so they have decided to organise their meals. The group will leave for Jaipur on Saturday early morning at 6.00 am. They will stop at Hotel Highway King for breakfast. The restaurant will charge INR 75 for a fixed menu. Around lunchtime, they will be close to their first stop Amber Fort. There are many eateries, and students can have lunch on their own. Finally, the group will have typical Rajasthani meals at Apna Jaipur resort for dinner. The special dinner will come at a special group price of Rs. 200 per Person. The published rate is INR 350. All the suppliers offer free meals for the TJG's guide.

Transportation

The company must evaluate the different modes of transport and the associated costs. First, there are intercity or to-and-fro costs to the destination, and then, there are costs of going around the place. Again, several options are available. Each segment transport cost must be identified and taken into account.

TJG advises the students to hire a dedicated 18-seater AC coach, which picks them up in Delhi and returns to the same place. The coach will also take them to various places in Jaipur. It is estimated that Delhi to Jaipur is 280 km. As per the itinerary, the group will travel around (less than) 700 km. Coach is available at INR 24 per km.

Details	Cost
Total mileage INR 24 X 700 km	INR 14,000
Tolls and parking	INR 1,860
Total	INR 15,860

Any extra kilometres over 700 km will be charged at INR 24 per km.

Tour leader/ Tourist Guide

The tour guide's cost – his wages and incidental expenses – must be factored into the total fixed costs. For example, the tour leader may engage step-on local guides at monuments. This expense also has to be included in the costs.

Rajesh Singh is the promoter of *The Jaipur Guides* (TJG). He is an accomplished guide who holds a Regional Level Tourist Guide license for northern India. As per the agreement between the Tourist Guide Federation of India (TGFI) and the Indian Association of Tour Operators (IATO), a tourist guide is entitled to INR 2,600 for full-day guiding. However, TJG offers curated experiences and has decided to charge Rs. 3,000 for full day's guiding. Rajesh accompanies the group for a day, and incidental expenses for him doubling up as tour leader is INR 700. (Total 3,700).

Miscellaneous expenses

Besides the above, many other expenses have to be included in the cost. These may include porterage or airport taxes, tolls, entrance fees at tourist attractions, among others. The company may also include some goodies for the tourist, and these must be included in the tour cost.

Toll taxes and parking have already been included in the transport cost. Other expenses are as follows:

Entrance fees at Amber Fort	INR 25
Entrance fees at Jantar Mantar	INR 50
Entrance fees at City Palace	INR 130
Entrance fees at Albert Hall	INR 40
Entrance fees at Hawa Mahal	INR 50
Total entrance fees per person	**INR 295**

Mark-up

Finally, the company must add a mark-up to the total cost to cover indirect costs and profit. The company gets a commission from suppliers who are part of the package and even those not formally parts of it, such as the restaurateurs, merchandisers, and others who have received the group. The highest indirect costs are the expenses of operating the company. A part of the cost has been apportioned to the package cost.

TJG has a policy of charging 5% of the total costs to contribute to the indirect costs and another 10% as the profit margin. The government of India taxes travel companies at 18% (GST).

What is included?

Inclusions: Transportation by 18-seater A/c coach, all tolls and parking, one breakfast and a special dinner, guide charges, entrances at sightseeing places,

Exclusion: Lunches, any shopping, drinks, porterage, etc.

Calculation of the total cost and arriving at the price

The price may be calculated as below:

		Description	Costs		
A	Meals			4,125	
	i.	Breakfast @INR 75 X 15 people	1,125		
	ii.	Special Dinner @INR 200 X 15 people	3,000		
B	Transport			15,860	
	i.	Coach 700 kms @ INR 24	14,000		
	ii.	Tolls and parking	1,860		
C	Tour guide			3,700	
D	Miscellaneous expenses @ INR 295 X 16 people			4,720	
E	Contingency expenses			1,595	
		Total (1)			30,000
F	Indirect cost @ 5% of (1)			1,500	
G	Profit @ 10% of (1)			3,000	
	Total (2)				34,500

So, the total cost of the package is INR 34,500 + GST for 15 people. The cost will be INR 2,714, say Rs. 2,725 per person. Total payable INR 40,875 – 25 % at the time of booking (non-refundable) and balance 75% at the start of the trip.

Factors affecting costing of the tour package

The pricing of the package is based on the cost-plus method in the above example. There are a few things that might affect the pricing of the package:

- Seasonality. Suppliers may change their price depending upon the season
- Foreign exchange rate fluctuations.
- Competitor's pricing strategies
- Inflation
- Changes in tax policies
- Changes in prices of suppliers
- Overheads of the company

Partners in business

Crowley & Karim (1995) defined partnering in business as a relationship of cooperative alliance established by players in a value chain. Partnering is, therefore, about trans-organisational cooperation for the realisation of a goal where participating entities stand to benefit. However, partnering is different from partnerships. While partnerships are legal arrangements, partnering is strategic coming together, generally, of smaller businesses. Tourism service providers at any destination are generally smaller players who create value for the visitors.

Tour guides are the most visible service providers who are often in extended contact with visitors. The human interface in tourism experience encounters is among the most critical touch-points in any tourism ecosystem.

While tour guides may have a defined limited role, tourists expect them to be friends, philosophers and guides at the destination. They approach tour guides for solutions to all their problems. It is, therefore, important that whether it is needed for foreign exchange, medication, unique experiences, access to non-public areas, shopping for conveniences, purchasing souvenirs, lost passport, missed train, food or country liquor, guests turn to tour guides. Guides must not disappoint. The job demands them to be the single window solution for all difficulties and anxieties – whether justified or not.

A trick of the trade is to build and sustain rapport with all possible stakeholders directly or obliquely connected to tourism – these are the partners in business.

Fellow tour guides

There will be other tour guides at any tourist destination who might have worked with you on various occasions. Many of them may have received training as guides along with you. You cross them as you work in the field. They have similar business interests and are, therefore, your competition. They face similar challenges and threats as you do. You must expect a long association with them, and together you can contribute to the development of local tourism standards and therefore the industry.

During the Covid 19 pandemic tour guides at every destination have felt similar pressures. They have come together to push for the revival of the industry and better working conditions. You must have a good rapport with fellow tour guides. Often they are a source of business for you. They might recommend your company when their hands are full. Experience suggests that many young guides have learned through socialising with senior guides, often turning them into mentors.

Trade associations

A natural next step to fellowship with other guides is having a formal trade association. Trade association raises issues concerning the guides' working conditions, often difficult for the individual to pursue. In addition, they lobby for guides' causes. During the Covid 19 pandemic tour guide trades associations have pleaded the case of guides and ensured benefits and protection. For example, resources were allocated to support the tour guides registered with the Registro Nacional de Turismo in Columbia. Similarly, in Israel, the ministry launched professional webinars and web-based courses to enhance the Israeli tourism industry's human capital, including tour guides (OECD, 2020).

Even during regular times, the associations negotiate with other bodies and public authorities for better working conditions for their members, including wages, fees, insurance, and other benefits. They also step in when any member runs into trouble. Associations are platforms for social and professional networking. These are portals for disseminating information and guidance on technical and procedural issues like legal and enforcement, and others.

It is advisable to obtain trade association membership at the earliest and abide by the association's code of professional conduct. Tour guide associations are also crucial stakeholders in maintaining professional standards.

Tourism companies

One of the key stakeholders in a tourism ecosystem are intermediaries like travel agents and tour operators. They attract and facilitate visitors' travel to a destination by designing and delivering tourism packages. It is here that they partner with other service providers, including tour guides, and are thus the prime source of business. On the other hand, guides are responsible for visitor satisfaction and are answerable to the tourism companies who have engaged them. The company is the customer while visitors are the customer's customers.

A beginner tour guide must visit the tour companies in the vicinity, introduce themselves, and present their strengths – language skills, experience, training, and ideas. They must maintain a profound connection with tour companies. When a guide leads a group of visitors, he/she is the face of the tour company from the party, and it is their primary duty to return the company's customers satisfied and happy. For this, they must understand the arrangements – what is offered and what is excluded. A sustained relationship with the company will let them know their limits regarding how far they can go, which the company will support. Guest spend a considerable amount of time with them, and they must ensure that customers are happy, entertained and satisfied. They must always present the company in good light.

On the other hand, they must keep the company updated on the group's movements and never break the communication. Always remain ready for the duty assigned, often at short notice, and oblige them when in crisis. Companies may sometimes require your services for familiarization trips. Provide the company with necessary feedback from guests and the destination stakeholders. Also, help tour companies improvise and design new products and packages.

Local tourism authority

Another important stakeholder is a local tourism authority. They are the face of tourism at any destination who provide direction by pulling all other stakeholders towards tourism goals. For tour guides, they are a significant source of business. Many visitors to the destination approach the tourism offices for inquiry and guidance. New tour operators also introduce them to local resources and service providers. They may refer guests to the guides and guides to tour operators.

Local tourism authorities often organise awareness and instructive programmes to educate local tourism businesses, including the guides, about destination image and related activities. For example, they organise workshops to educate about new product launches and future developments. Often they are responsible for organising the recruitment of tour guides and their skill training programmes. They may also certify or license the tour guides.

If you want to operate as a tour guide, the first thing is to register with the local tourism authority. Even if it is not mandatory, you must visit them and introduce yourself. Leave your contact details – your cards or brochures – with them. Visit the authority and meet them frequently so that you are on top of their recall. Remind and request them to refer business to you. Often the authorities would assign visitors to you. Do not turn down their requests to accompany groups. Contribute to making the local tourism authority successful. Attend workshops and meetings and provide feedback and constructive criticism to improve visitor satisfaction.

Hotels and alternative accommodation providers

Every evening guests will retire to hotels or alternative accommodation. Accommodation is an essential component of any tourist destination. Often tour guides are requested to help with transfers – to pick guests from the railway station/airport/seaport, lead them to hotels, and help them settle in comfortably. Other times, guests staying with these properties will seek guides to take them around. Accommodation establishments are, therefore, a source of business.

Guides must visit the local hotels and develop a good rapport with them, especially the front office staff – the counter, travel desk, and concierge staff. Tour guides may leave their details with the front office/ travel desk.

Before a group is received to be transferred to the hotel, guides may visit the property and learn about the facilities and operations details like parking, restaurant, lifts, evacuation, facilities in the room, door locking system, mini-bar policy, room service, hot and cold water facilities, among other details. When the tour guide arrives with a group, they may help the front office with rooming lists and guests' details (passports, identity, etc.). They may also help guests with the hotels and neighbourhood details and even allocate rooms.

The hotel gives you business, and you may return the favours by spreading the good word about the property. You may also receive incentives and discounts from hotels for referring them.

Restaurants and eateries

Guests will often ask for a reference for a good restaurant or a speciality restaurant. Tour guides must be in a position to advise. Restaurants and eateries may be part of the itinerary, and the guide may accompany the group for lunch or snacks as per the schedule. A good rapport with the eatery will ensure that your party is served on priority without waiting too long. For unscheduled recommended visits by your guests to eateries, they may be willing to return favours for recommending them.

Tour guides should know about the city's restaurants, their specialities, and their clientele and price bands. They must first visit the restaurants and similar outlets and introduce themselves. It is good to know the owners, managers and the chef personally. The tour guide often has to negotiate time slots for their party and keep the restaurant informed about the party's arrival to prepare in advance to get preferential treatment. Sometimes, the guide may have to talk to the chef to explain the guests' expectations or requirements. Similarly, guests may be advised about an eating joint based on the time, budget and preference.

Local transport providers

Guests frequently require local transport like cabs, autos, rickshaws. They may ask the tour guide to organise transportation to help with local sightseeing/ shopping on their own – or just a joyride around the city. Sometimes they prefer an additional sightseeing trip to a nearby place. Sometimes, an independnet traveller may ask the local transporter to arrange a guide.

It is, therefore, important for tour guides to be in touch with reliable local transport companies. In some countries, the local transport is quite organised, and in those cases, the tour guide may only advise the clients. The tour guide may patronise a few taxi owner-drivers, auto-drivers, rickshaw-pullers, tongas, and others where the local transport is not much organised. The guide may have their contact details. Within minutes of the guest asking for help, the tour guide should arrange a service provider for a ride for guests!

Authorities at sites

A destination may have attractions like museums, archaeological sites, galleries, shows, live performances, and religious places, among many others, where the guide leads the guests. A guide must be familiar with the places and their protocols. They may have to procure the necessary permissions for the group's visits. Guides may have to talk to the responsible person at the authority to know about convenient time slots for bringing the group wherein the group gets adequate attention. Guides must also remain aware of special days, events, initiatives, restrictions, etc. The authorities may also help the guide with important information about the attraction.

Guides must visit these attractions and introduce themselves to the responsible persons like curators, administrators, priests, etc. Leave your card and brochure. Oblige them as and when possible, and promote visits to these sites. When at the site, help the authorities maintain decorum and organise the show.

Local civic administration

These are also essential stakeholders at any destination. These may include the city administration, police and others. They are responsible for broad discipline, maintenance of infrastructure and sometimes operations of the tourist sites. They also control permissions for visits and sometimes issue licences (for local guides). Nevertheless, tourism is their interest, and they are happy to support responsible tourism.

A tour guide must approach the local administration to introduce himself/herself. They must seek necessary permissions wherever necessary and keep the authorities in the loop about their activities. Guides must always help and support the local civic authorities enforce law and discipline by encouraging guests to follow their advice. Guides are on the field, and they must inform authorities of any untoward event or suspicious activities.

Shopping establishments

Shopping is an essential component of any travel. Most visitors like to carry back local stuff as souvenirs. Tourist guides often receive commissions on the purchase made by patrons. Therefore, it is an oft cursed part of a tour guide's role when they are accused of pressuring guests to buy from chosen outlets, often at a higher price. Many destinations have a code of conduct that discourages tour guides and tour leaders from accompanying the guests to shopping establishments. However, guests will ask for references. They want to be sure of the purchase. Shops often return the favour for recommending them. Therefore, commissions from shopping are a significant revenue stream for tour guides.

Self-imposed discretion is important. Never leave your guest vulnerable to cheating, something they are sceptical about. A cheated guest will have significant negative word-of-mouth for the destination – and part of the blame will be on the tour guide. Always provide expert and sound advice. Learn about the opening and closing times of markets and shops. Know who is a credible seller and what are the correct prices. Can prices be bargained? Guides should refer to only good shopping establishments where reasonable prices and quality are assured. Guides should build rapport with such establishments – leave your visiting card and obtain their contact details.

Other service providers

Many sundry service providers add to visitors' experience – porters, hawkers, performers, jugglers, priests, and many others. Often a situation may emerge where the guests may demand support or entertainment. For example, in South Asia, guests may require help with transferring their luggage at a train or bus station; or they may want to witness some services or rituals at religious sites. If a guide can arrange for such support and services at short notice, this creates an excellent impression on the guests.

On the other hand, these service providers get business and are willing to oblige. A guide's job is to get the right persons at the right time, for example, proactive porters as the guests struggle with their luggage. So, a good rapport with them is always good. The guide may offer tips to these providers to keep them in good humour. There are examples of local service providers like security guards, caretakers and even hawkers displaying extra courtesies (making guests feel special) as they have been tipped; this pleasantly amuses them.

Community

The local community is one of the essential contextual elements of any tourism ecosystem. It provides a vital backdrop for the unique experience of the tour. Subtly, all community members play their role in staging an experience for the guests. In a way, they are the agents of tourism experiences. However, they may be obstructed and offended by visitors' movements. In recent years this has manifested itself as 'overtourism'.

Tour guides need to engage extensively with the community. Guides should educate the community about the benefits of tourism and encourage them to welcome the visitors and treat them well. The tour guides' responsibility is to organise their visits and movement through the community in the least obtrusive manner. They must also educate the guests about local sensitivities and appropriate behaviour, including dress and how guests conduct themselves.

Family

Last but not least is the family. They allow guides the time that was due to them. They accommodate your awkward arrivals and departures, meetings, and the accompanying unknown people. They provide emotional and moral support to the tour guide.

It is therefore essential that guides keep their family members in good humour. It is suggested that guides frequently talk to their family and tell them about their work and job descriptions. In addition, they must remain in touch with the family while on tours and frequently express their gratitude.

Thus, guides should understand the importance of partners in business and network extensively with them. Be good and invest consciously in relationships.

Legal considerations

Tourism is a significant economic activity that employs many people and is dependent on many sectors and industries. It is essentially an organic ecosystem of interdependencies. While market forces are at play in contemporary society, a legislative framework is often required for even-handedness among interdependent interests. A visit by a tourist is the basic building block of this ecosystem. Unfortunately, tourists visiting unknown destinations are often unaware of how the law of the land plays out their rights and responsibilities and the deterring punitive framework in place. The closest advisors of a visitor at a destination are the tour guide or the tour manager. Therefore, a professional tour guide must be aware of the laws related to tourism-related activities. They must also be aware of laws related to their business and operations.

Three types of issues might affect the operations of the guides within the legal framework of the land.

a) Laws governing the recruitment, licensing, registration, and operations of tour guides;

b) Laws related to conducting tour guiding as a business. For example, contract law, consumer law, employment law, fair trading law, property law, insurance law, the law of agency, etc., depending upon the country;

c) Laws related to the visitors' activities at a destination. These may include laws related to foreigners, passports, imports-exports, currency, restrictions for foreigners; laws related to preserving heritage or wildlife or natural areas or public order, etc.

Travel and tourism laws

Given the significance of tourism in society and economy, different countries and regions have enacted various laws. Some have comprehensive laws for tourism by name, while others have selectively approached tourism legislation. While the earlier have tried to address multiple issues, the latter has addressed specific issues. These laws regulate, permit, encourage, empower or restrict the leisure and commercial activities of visitors and the service providers, for example, by recognising the rights of the tourists vis-à-vis the industry segments.

There are seven basic tenets of tourism laws. These include i) Travel is a legal right; ii) Reliable and safe transportation must be readily available; iii) Safe and adequate accommodations must await the traveller; iv) All travellers should have access to such accommodations; v) Travel and accommodation costs must be reasonable; vi) Regulation of the travel and tourism industry is necessary; and vii) Redressal mechanism for transgressions of rights and regulations is necessary.

Tour guiding laws

Given the importance of tour guides in the tourism setup of a destination, many countries have enacted laws that regulate the administration of tour guide operations. For example, in India, the regulations for tour guides are derived from Rule 8(d) of Ancient Monuments and Archaeological Sites and Remains Rules, 1959. These regulations standardise tour-guiding activities by specifying the rights and obligations of tour guides and protecting and promoting healthy tourism that includes fairness and satisfaction of visitors.

These rules often prescribe a uniform qualification, training and examination for certifying or licensing tour guides as in China and India. They may also identify the agencies entrusted with these responsibilities. In addition, these rules sometimes prescribe retraining and upgradation of tour guides by linking the same with the renewal of licenses, thus setting the validity of a certificate or license. Sometimes these rules also outline the instrument for determining the compensation, perks and benefits available to tour guides in various situations. Often these rules also include a code of conduct for the tour guides.

Guides must understand these laws as they directly bear on their operations.

Corporate law

Corporate law is also called business law or enterprise law, or in some countries like India, company law. It is a set of civil laws governing the rights, relations, and conduct involved with forming, owning, operating, and managing a corporation. Corporate law deals with matters related to the life-cycle of a corporation. It thus regulates the formation, financing, governance, and death of a corporation. In many countries, guides form and operate tour guiding companies of one or more partners and employees. Broadly, corporate law deals with five dimensions of a company's existence – a separate legal personality, limited liability, transferable shares if it is a public company (Note: private companies can also have the transfer of shares), delegated management under a broad structure, and investor ownership.

Given the provisions of the local corporate regulations, tour guides may decide upon the form of the company that they would like to create. Guides in India, as discussed earlier, can now form a one-person company (OPC).

Agency principle

This is not a separate branch of law but refers to the *fiduciary* relationship principle present in other streams of laws. A tour guide must be aware of this principle. Guides typically represent a travel agent or a tour operator. They are thus the company's agent, while the company is the 'principal'. In most jurisdictions, this is covered under fiduciary relationship principles, which deals with a set of contractual, quasi-contractual and non-contractual relationships between agent and principal. The relationship often exists by express consent but can also be found by the parties' conduct.

A travel agent often depends on a tour guide to conduct business with the guests whereby the travel company expressly or implicitly, authorises the agent, the tour guide, to work under their control and on their behalf. The guests may rely in good faith on the representation by tour guides who present themselves as representatives of the tour company. Thus, if it is subsequently found that the tour guide was operating without necessary authority, the guide will generally be held liable. Whereas the tour company can be held liable only for the outcome of actions that they direct the tour guides to perform.

Contract law

A tour guide often works for a travel company as a freelancer who engages them through a contract that requires them to conduct themselves in a particular way. Further, the successful conduct of a tour often involves several partners of the travel company who might be tied to each other via a contract. A contract is a promise enforceable by law. Such a promise may be to do something or refrain from doing something. Legal complications arise when one of the two consenting parties fail to perform the contractual legal obligation. In such cases, the aggrieved party can sue for damages or ask the court to direct the other party to perform as promised.

A contractual relationship begins with an offer which is a proposal by one person, who expresses their willingness to enter into a contractual obligation in return for a promise, act or forbearance. A contract emerges from accepting an offer, which assures the offeree of an offer. In Anglo-American common law, the formation of a contract generally requires an offer, acceptance, consideration, and mutual intent to be bound. However, not all agreements are contractual, as the parties generally must be deemed to have an intention to be legally bound.

Fair trading laws

Fair trading laws protect consumers and businesses against being misled or treated unfairly by traders or shops. These laws check misleading and deceptive conduct, unsubstantiated claims, false representations, and unfair practices. In addition, these laws help businesses operate fairly and competitively and make sure their customers are adequately informed and protected. hey may also provide additional protection for the disadvantaged (Greenwood & Dwyer, 2015). Competition laws and consumer laws are examples of fair trading laws.

If a visitor goes back dissatisfied, it adversely affects the destination's competitiveness. Thus destinations need laws to protect consumers, and a visitor who hires the services of a guide through a travel agency is a consumer of both.

Consumer protection or consumer law looks into the legal relationships between individual consumers and the businesses that sell goods and services to them. It addresses product liability, privacy rights, unfair business practices, fraud, misrepresentation, and other consumer and business interactions. More than fifty countries have some consumer laws that ensure consumers receive goods and services of reasonable quality, requisite information about the product, surety about consumers' health, safety and economic interests; and some compensation for loss or injury arising from defective goods or services.

There may be some confusion regarding the territoriality of the complaint, but in most cases a consumer has to apply local consumer laws to seek restitution for a faulty service or product.

Guides must understand how they are liable for their actions that are deficient or cause harm to their consumers, the tourists.

Employment laws

In some cases, the tour guides and tour leaders are employed by the travel agent or the tour operator. Such employment may be governed by the applicable employment law, if any. Also referred to as labour laws, these regulate what employers can expect from employees and can ask them to do, and the employees' rights at work. These laws may also cover what an employer must pay the employee for their work, and in some cases, they set the minimum working conditions for employees. Thus, these laws cover minimum wages, working hours and overtime pay, family and medical leave, collective bargaining, safe working conditions, discrimination, dismissal, and child labour.

Insurance law

There are risks while travelling to a foreign place. These risks include those associated with transport, bodily injuries, risk in the process of accommodation and other types. As a result, it has become commonplace to obtain and maintain appropriate insurance coverage to manage risk while travelling. Insurance is a contract in which one party, the 'insured', pays a premium, and the other party promises to reimburse the first in case of certain types of losses, like illness, property damage, or death. Insurance, thus, transfers the financial risks to a third party – the insurance company.

There are compulsory and voluntary insurances. The things travel insurances cover include some of the following: tourists' health and their goods; risks for travel agencies; outbound tourists for travel abroad; inbound foreign tourists; civil liability for owners of premises and of means of transport; injury, accident and other unforeseen circumstances; and payment of medical bills.

Insurance laws are a set of laws that deal with insurance contracts, policies, and claims. They cover the regulation of insurance business, regulation of the content of insurance policies, especially consumer policies, and regulation for handling claims. A tour guide must be aware of the application of these laws to visitors, travel agencies and their operations, and themselves.

Intellectual property law

While it appears that intellectual property (IP) has little to do with tourism, the converse is true. As tourism grows, it gets competitive, and it becomes necessary to provide it with a cover of intellectual property laws. IP can be broadly classified into an industrial property like patents, trade secrets, trademarks, industrial designs and geographical indications; and copyright covering literary works, films, music, artistic works like paintings, photographs and sculptures, and architectural design. Books, brochures and web pages are also covered. These may help create a distinctive identity, promote local culture and heritage, and add a revenue stream that would otherwise have not been possible.

Protecting intellectual property is a crucial endogenous development tool for disadvantaged communities that create customised products and services for differentiated value-added experiences associated with tangible and intangible community assets for the visitors. Guides must be aware that their company and their guests do not violate intellectual property in any way.

Environmental laws

The environment is a top concern for the planet. These concerns include climate change, rising air and water pollution, deforestation, land and soil degradation, and waste management. Mitigation of adverse impact is a priority in every sphere of human activity. As tourism grows, stakeholders become increasingly aware of its impact on the destination and climate change. Much tourism is contexted in its natural surroundings. Therefore, habitat preservation is ensured through a legislative framework, especially the environmental laws. Tour guides must be aware of applicable laws and ensure that visitors do not violate the same.

Environmental law refers to a set of laws that protect the environment. These include a legislative framework of principles, policies, directives, and regulations enacted and enforced by local, national, or international entities to regulate human activity and its impact on the nonhuman world. In addition, they include acts enacted for conservation, preservation, and protection of flora and fauna. Many places have acts related to wildlife conservation and protection, forest conservation and protection, pollution prevention, protected areas regulations, and many others.

Heritage protection laws

Cultural heritage is a crucial travel motivation for the visitors. Cultural heritage includes the material and immaterial assets which are evidence of the human creation in the past and the present or as shared works of humans and nature.

These are considered significant due to their archaeological, ethnological, historical, artistic, architectonic, urban, ambiental, technical, sociological and other scientific or cultural values, characteristics, contents or functions.

Concern for the protection of heritage is not recent. On the contrary, humans all through their history have shown a desire to conserve their heritage. The first evidence available is from the royal edicts of Alexander and later Romans in the Mediterranean Basin. They recognised that physical materials from the past represented a record of cultural heritage for subsequent generations. This sentiment was echoed by proclamations of Caliphas in the 8th-century Middle East, and in Imperial China of the 11th century. Subsequently, Sweden in 1666 and Russia in 1704 had similar diktats. Finally, Europe and colonial India began to enact heritage protection legislation in the 19th century.

Heritage is thus protected by laws in most parts of the world. For example, the Indian constitution under Article 49 makes it obligatory for the state to protect every monument or place or object of artistic or historical interest as declared by or under law made by Parliament to be of national importance, from spoliation, disfigurement, destruction, removal, disposal or export. India has relevant Acts dating back to British times like the Ancient Monuments Preservation Act of 1904, the Treasure Trove Act of 1878, the Ancient Monuments and Archaeological Sites Remains Acts of 1958 and the Antiquities and Art Treasures Acts of 1972, among many others. Every destination is differently placed and has unique concerns. They enact laws accordingly. For example, in Australia, there are laws to safeguard its natural and cultural heritage – Australia's world heritage properties, national heritage places, wetlands of international importance, migratory species, listed threatened and ecological communities, and the Commonwealth marine areas, the Great Barrier Reef Marine Park among others. A variant of these are cultural property laws that protect and regulate the disposition of culturally significant material (Sullivan, 2016), including historical real property, ancient and historical artefacts, artwork, and intangible cultural property.

The tour guide must be aware of these acts and the rules framed by these acts. These are the basis for dos and don'ts for visitors at heritage sites.

Immigration laws

Immigration law refers to statutes, regulations, and legal frameworks regulating immigration into and deportation from a country. Strictly speaking, it is distinct from other matters such as naturalisation and citizenship, although they are sometimes amalgamated, as in the US.

Citizenship confers the right to a citizen to obtain the passport of the dominion. These laws also determine the types of passports that are issued. Also, when a foreign national enters a dominion without permission, overstay their visit, or lose their legal status for various reasons, immigration law determines the detention and removal proceedings. The social and political environment of a place has a significant bearing on immigration laws.

Many countries conflate immigration with policies and regulations for visas. Visa is a limited authorisation granted by a state to a bearer to enter, remain within, or leave its territory. The state decides the types of visas to be granted like tourist visas, business visas, student visas, medical visas, or more lately, digital nomad visas, among many other types. They also decide about the costs, conditions and the process of obtaining such visas.

A tour guide must understand these laws so that they know what visitors can do, the validity of visas, or, as in some cases, reporting of arrival and activities of visitors. They may also be required to advise guests on the loss of passports.

One of the critical conditionalities could be restrictions on visiting certain fragile, protected, and disputed areas within the country being visited. In some cases, foreigners must register with local authorities. Tour guides must be aware of such restrictions and the processes to facilitate the visitors.

Customs law

The Customs authority of a country is responsible for collecting tariffs and controlling the flow of goods. They, thus, levy duties for bringing in and taking out goods. Tourists are typically affected as they routinely carry some goods and souvenirs with them. Each country has its customs laws and regulations enforced by their respective customs authorities. Tourists, as bearers of goods, must know that some goods may be restricted or forbidden entirely. When required, the tour guides must advise guests about customs regulations. For example, while shopping, guests are often unsure if they can carry a souvenir or something else out of the country. For example, they may like to know if they can take shells and corals from the country.

Many countries have created the red and green channels convention to facilitate international travellers at their ports. Passengers who have nothing to declare, that is, those who carry goods within the permitted customs limit and that are not prohibited, may pass through the green channel. Others must pass through the red channel and declare their goods. AT EU airports, for intra-EU travellers, there is a blue channel also. It is for arrivals from non-EU countries via an EU country, for example, Delhi-Paris-Amsterdam, and if the baggage was cleared from customs at the first port of arrival in the EU.

Health laws

Public health is a growing concern for governments around the world and communities are concerned about their health and hygiene issues. Government at different levels, federal, state, or local, creates a legal framework including laws, rules or regulations to delineate and regulate the rights and responsibilities of authorities, health workers, companies, civil society and the general populace.

The management of tourists' well-being as guests, and their impact on the health of the host community are concerns. Health and related laws address these issues, and a tour guide must be aware of the same. For example, in a post Covid era, vaccination and travel are conflated. Even earlier, countries would require visitors from some parts of the world with pandemic history to carry health clearances or vaccination certifications.

Health laws also impact medical and wellness tourism. Certain countries may restrict specific medical procedures – for example, surrogacy is not permitted by Indian laws. Often, alternative medical procedures are practised in many communities, and may attract and excite visitors. A tour guide must understand the legal implications of liability (Consumer protection) before advising or allowing any quasi-medical treatment.

Other laws

Every place is uniquely situated. History, geography, legacy and contemporary society and regulations, among other things, shape it. Leadership has a vision for the place, implemented through a legal framework. Thus, every destination has its set of instruments that ensure compliance and desired behaviour from both the hosts and the guests. Tour guides, the representatives of the destination and professionals who know and interpret the local laws for the visitors must ensure that guests understand such laws and regulations and comply.

For example, chewing gums are prohibited in Singapore; the Cigarettes and Other Tobacco Products Act, 2003 (COTPA) bans smoking in public places in India; alcohol is controlled in many countries, especially the Islamic countries – in some cases, like Libya, Saudi Arabia, Kuwait, it is completely banned, whereas it is banned for Muslims in Maldives, Iraq, Qatar, and Pakistan; there is a legal age below which guests cannot be served alcohol in many countries; Belgium has recently banned halal and kosher meat.

There are a variety of laws covering some recreations too. For example, while there is extensive gambling in India, it is illegal. Similarly, a statute restrict gambling in Singapore – only integrated resorts like Marina Bay Sands and Resorts World Sentosa can operate casinos. Similarly, different countries have taken a different position on recreational hunting. For example, Costa Rica has banned all sport hunting and trapping inside and outside protected areas. Botswana, Zambia, Kenya, Brazil, Singapore and many other countries also have restrictions on recreational hunting. Convention on International Trade in Endangered Species of Wild Fauna and Flora (CITES) formulates a country-wise annual quota for hunting and export of selected species. While India prohibits trophy hunting, neighbouring Pakistan has issued special permits to Arab royalty selectively to hunt the endangered bird species Houbara Bustards in allotted hunting areas spread over Sindh, Balochistan and Punjab provinces. The hunters can hunt the permitted number of Houbara Bustards in a limited time safari.

Visitors rely on tour guides to educate them on local laws and restrictions and help them comply. Therefore, tour guides must be aware of laws and translate them into dos and don'ts for their guests.

Ethics and conduct

Ethics is the moral philosophy of systematising, defending, and recommending right and wrong behaviour concepts (IEP, 1995). Though ethics are difficult to pin down, they tend to prescribe what humans ought to do, usually in terms of rights, obligations, benefits to society, fairness, or specific virtues (Velasquez et al., 2010). Tour guides often face a dilemma about right or wrong as they carry out the assigned roles. Two terms, 'code of ethics' and 'code of conduct' are often used interchangeably, but there is a difference between the two. 'Code of ethics' is a broader term reflecting on the company's values to transact business fairly, considering the obligations to the stakeholders. The code of conduct, however, is individual. It aims to monitor and restrict individual behaviour, in our case, the tour guides. It is focused more on rules and compliance than the underlying principle. Companies tend to combine the two codes into one for practicality.

However, a mere code of ethics cannot guarantee ethical behaviour. It only sets forth values, ethical principles, and ethical standards. A tour guide's ethical behaviour should result from their commitment to ethical practice. Often destination leadership and the companies they represent prescribe a code of conduct for tour guides. For example, the Travel Industry Council (TIC) of Hong Kong, or the Battlefield Region Guides Association of Kwazulu Natal, prescribes a code of conduct for their tour guides.

General principles of ethical discussion

Guides are placed in a situation where they must address multiple stakeholders. Even while they work independently or on behalf of the companies, they must be responsible for these stakeholders, referred to as 4Cs: Customers and clients, Community, Company and the Context. The discussion on ethics and ensuing codes for conduct revolve around a few established principles (IEP, 1995).

a) Duties towards the clients and customers

b) Duties towards the community

c) Duties towards the Company

d) Duties towards the Context – the destination, leadership, and governance.

Principle of beneficence

Also called the principle of service, this refers to helping those who need it; in this case, the visitors with their stay and immersion in the local experiences. Tour guides are expected to put their service to guests above their own interest. Tour guides leverage their knowledge, values, and skills to help guests in need. They must also go beyond the call for pro-bono volunteering to help guests without expecting a significant financial return.

> *Tourist guides shall strive to provide visitors with the highest standard of service in accordance with the provisions of the travel service contract and the itinerary. The itinerary shall not be altered without the consent of the visitors and the travel agent concerned. If the alteration is necessary because of emergencies or special circumstances, tourist guides shall seek approval from their travel agent and clearly explain the reasons for such alteration to the visitors.* (TIC, 2010)

Principle of nonmaleficence

This principle says do not harm others. However, this can be extended to mean that where harm cannot be avoided, guides must minimise the harm to the guests, and must take all preventive action not to put guests at risk. A corollary to this principle is that it is wrong to waste resources that could be used for good. In all, the principle suggests that every action of the tour guide must produce more good than harm.

> *Tourist guides shall always attach top priority to the safety of visitors and remind them to be aware of their personal safety.* (TIC, 2010)

> *In case of any accidents or visitors feeling unwell, tourist guides shall help arrange immediate treatment for the visitors and seek assistance from their travel agent without delay. They shall report the case to the police if necessary. A corollary to this principle is to ensure that no harm is caused to the destination also.*

> *To protect the reputation of tourism in our country by making every endeavour to ensure that guided groups treat with respect the environment, wildlife, sights and monuments, and also local customs and sensitivities.* (WFTGA, 2006)

Principle of honesty

It is expected that tour guides conduct themselves trustworthy and not deceive others. They are honest and truthful in all their dealings, and they do not deliberately mislead or deceive others by misrepresentations, overstatements, partial truths, selective omissions, or any other means. Tour guides should act honestly and responsibly and promote ethical practices in the company they represent.

> *To ensure that as far as possible, what is presented as fact is true and that a clear distinction is made between this truth and stories, legends, traditions, or opinions.* (WFTGA, 2006)

A member of the Guild should not abuse complimentary meal privileges offered by food establishments and does not solicit gratuities. (The Guild, 2017)

Principle of integrity

That the tour guides display personal integrity and the courage of their convictions by doing what they think is right even when there is tremendous pressure or allurement to do otherwise

Tourist guides shall avoid situations which may lead to an actual or perceived conflict of interest situation and report any unavoidable conflict of interest situations to their travel agents.

(For example) a tourist guide has shareholdings either by himself/herself or his / her immediate family member(s) in another travel agent which competes with his / her travel agent; and

(For example) a tourist guide is of close personal relationship with a retail shop with which he/she has business dealings on behalf of his / her travel agent. (TIC, 2010)

Principle of lawfulness

Do not violate the laws and abide by the rules and regulations at the destination. Guides should help visitors understand the law and comply with it.

The guide will abide by the existing rules and any other rules and conditions that may be framed from time to time by the Ministry of Tourism and the Archaeological Survey of India, failing which the identity card issued to him/her shall be withdrawn. (Chowdhary, 2014)

Tourist guides shall abide by and help visitors to understand and abide by the laws of Hong Kong. (TIC, 2010)

Principle of autonomy

This acknowledges a person's right to the decisions concerning their own lives. Also called the principle of human dignity (Gracyk, 2012), this suggests that a tour guide should not interfere with the decisions of competent adult guests.

Tourist guides shall ensure that visitors understand their consumer rights, which include:
-the right to make or not to make purchases,
-the right to obtain information on products and services,
-the freedom to choose products and services, and
-the right to complain request their purchases to be exchanged or refunded.
(TIC, 2010)

Principle of competence

It requires that the tour guides practice within the areas of their operations and develop and enhance their professional expertise. They must participate continuously in training to enhance their capabilities.

A member of the Guild provides a skilled presentation of knowledge, interprets and highlights surroundings, informs, and maintains objectivity and enthusiasm in an engaging manner. (The Guild, 2017)

Principle of justice

This acknowledges that a person should get what they are owed or deserve. Thus guests should be treated equally and impartially. A natural corollary to this is that guests should not be unfairly or unduly burdened, for example, pushing them for unnecessary shopping or to donate to causes.

The guide would not canvass business of any kind on behalf of the business houses, travel agents, hoteliers, paying guesthouses, shopkeepers, transport operators, etc., nor he/she would accept monetary commission from them or indulge in any other unethical practices. (Mizoram Tourism, 2020).

In no way discriminate in rendering service to any tourist on any basis, e.g. colour, gender, ethnicity, nationality, physical challenge, age, etc. (Gauteng Guide Association, 2000)

Principles of professional compliance

This advises tour guides to comply at all times with all requirements – mandatory and expected. Mandatory requirements like leading a group only where they are authorised; or carrying the licence or wearing a badgeswhile guiding. The expected requirements are that the guides present themselves professionally – well-groomed, polite, forthcoming, etc.

Wear the appropriate tourist guide badge and will carry his/her registration card. (Gauteng Guide Association, 2000)

A member of the Guild maintains good personal hygiene and uses proper judgment in dressing appropriately for all tours. (The Guild, 2017)

Principle of professional conduct

This relates to how tour guides conduct the business. In addition, the principle will often address the local concerns.

The guides will charge guide fees to be fixed from time to time by the Tourist Guides Federation of India in consultation with the Indian Association of Tour Operators. (Chowdhary, 2014)

A member of the Guild accepts each tour as a serious commitment and cancels only when he or she is physically unable to perform the work; under such circumstances, he or she shall provide as much advance notice to the client as possible. (The Guild, 2017).

Code of ethics for tourist guides

(Ministry of Tourism, Government of India)

The Tourist Guide shall always be aware of projecting a good and positive image of India adhering to the ancient Indian concept of 'Atithi Devo Bhava[2]'.

The Tourist Guide shall consider himself a friend of the tourists, be courteous and helpful, never betray the trust placed in them by the tourists and carry out his ethical and moral responsibilities with dignity.

The Tourist Guide will, therefore, bear in mind that 'word of mouth' publicity and comments have a tremendous impact on the creation of an image for the country. This is several times more valuable than the most expensive advertisement that can be put out.

The Tourist Guide will always be courteous, efficient, tactful, kind, understanding and project the warm and hospitable nature of Indians.

The Tourist Guide will ensure that he will give correct and all possible information of the latest developments in the country and keep himself up to date. He will understand and convey the positive aspects of his culture and tradition and be well-versed in matters relating to local customs, habits and traditions to ensure that both the tourists and the local people respect each other.

The Tourist Guide will always be punctual, well-groomed, presentably dressed, well mannered, sober and cultured. His dedication and conduct should ensure that the image of the profession or that of the country is never tarnished.

The Tourist Guide shall at all times be reliable with the entrusted duties and be of exemplary conduct, not only in dealing with tourists but also with others such as Govt. Officers and members of the public.

The Tourist Guide shall refrain from unethical or discriminatory conduct and deal with proprietary and impeccable honesty. He will refrain from entering into shoddy deals for monetary and other gains, nor will he intentionally mislead the tourists.

The Tourist Guide shall refrain himself from all possible arguments and remember that the customer is always right.

The Tourist Guide shall uphold his/her status and reputation with a view to fomenting closer cooperation, understanding and collaboration with his/her counterparts throughout India.

The Tourist Guide will respect and safeguard our heritage, monuments and ecology

2 *Athithi Devo Bhava is a Sanskrit phrase meaning "Guest is God".*

Key points

A tour guide business could be a one or two-person initiative, or a more prominent company that offers tour guiding services and curated tours. This chapter focuses on a few nuances of operating a tour guiding company. The first part discusses setting one up. One should begin with checking the requirements with the local authorities. The next step is market research to zero in on a target market. After that, a suitable type of company should be formed, and a name be registered. Also, the entrepreneur guide must work out the startup costs.

Marketing the company and the products is the next step. Initially, the business should be straightforward. The entrepreneur should also have some training in keeping personal and business accounts. They must prepare a detailed business plan that serves as a roadmap for the future. The critical components of a business plan are executive summary, market analysis, competitor analysis, sales forecast, company management, operations, staffing and projected financial statements, especially the balance sheet and profit and loss statement.

An itinerary is an essential product for a tour guide. Some of the principles to be kept in mind while designing an itinerary include: downtime, flexibility, and pleasure. The itinerary must be formatted to be visually appealing. It is essential to competitively price an itinerary, and that will require determination of the costs: fixed, variable and variable-fixed costs. Some of the essential components of the cost of an itinerary are: the meals, the transportation, tour guide/ tour leader costs, and other miscellaneous costs. After that, a mark up is added.

It is necessary that tour guides identify the key partners in their business and appreciate their role and contribution. These partners are: fellow tour guides, trade associations, tourism companies, local tourism authorities, hotels and alternative accommodation providers, restaurants and eateries, local transport providers, authorities at the sites, local civic administration, shopping establishments, other service providers, the community and the tour guide's family.

Tour guides must be aware of the legal framework and local regulations. Visitors rely on tour guides to educate them and deal with the local laws. Broadly there are three types of legal issues: one, the legislation governing the recruitment, licensing, registration, and operations of tour guides; two, the legislations related to activities of visitors to a destination; and, three, the legislation related to conducting tour guiding as a business.

Finally, the chapter deals with ethical conduct issues related to the clients and customers, and the duties of the tour guide towards the community, the company, and the context. The context includes the destination, its leadership, and governance. The chapter also deals with some principles that form the basis for ethical conduct.

Key terms

Business plan	A formal written document that details the goals of a business, how it will go about realising those goals, and the time frame for achieving them. It is a firm's go-to-market plan, based on its vision and mission, business purpose, and market research. It also speaks about human resources, operational, managerial strategy and financial projections.
Code of conduct	A set of rules outlining the norms, rules, and responsibilities for the professional conduct of an individual. It lays out the principles of behaviour, standards, and the moral and ethical expectations from this profession.
Company	A legal entity formed by a group of individuals to engage in and operate a business enterprise to achieve specific objectives.
Consumer law	Or consumer protection laws, are the regulations and statutes that protect the consumers of products and services against unfair practices of the suppliers.
Contract law	A set of regulations and statutes that address issues involving contracts, which are legally enforceable agreements between two or more parties.
Corporate law	Also known as business law or enterprise law or sometimes company law, is the set of laws that includes the rules, practices, and regulations that deal with forming, owning, operating, and managing a corporation.
Ethics	A system of moral principles based on the premise of right or wrong, manifest in terms of rights, obligations, benefits to society, fairness, or specific virtues.
Fixed costs	Sometimes referred to as indirect costs, are costs that do not change with an increase or decrease in sales or production volumes. Examples of fixed costs are rent, salaries, insurance, depreciation, and interest.
Immigration laws	The framework of statutes, rules, regulations, and legal precedents that deal with who can legally enter and remain in the country. Academically speaking, it does not deal with naturalisation and citizenship, but the two are often conflated.
Intellectual property laws	Statutes, rules, regulations, and legal precedents that protect the rights of creators and owners of inventions, music, writings, design and other intellectual works. Instruments for protecting IP include copyrights, trademarks and patents.
Itinerary	A detailed plan for a journey outlining the scheduled events, including places to be visited at specified times, means of transportation to move between those places of visit, activities at those places and stay details at those places.
Law	Rules and guidelines set up to govern behaviour to maintain order and comply with norms. Laws determine what an individual can do or cannot do.
Legislation	Statutory law, i.e. a law passed or enacted by the legislature of any authorised body of the state. Sometimes also refers to the process of enacting the law.
Market research	An integral part of business planning, is the systematic process of collecting information about a target market, consumers, competitors and the industry as a whole, generally to evaluate the feasibility of a new product or service.
Regulation	Rules made by an authority to limit, steer, or otherwise control the way something is done or the way people behave. They are different from laws in that these are standards and rules adopted by administrative agencies to enforce laws.
Variable costs	Vary as the quantity of the good or service that a business produces changes, i.e. depending on the volume of activity. Examples include tickets or meals of guests.

Discussion questions

1. Visit the local administration office to understand how can you set up a tour guiding company.
2. Visit the local attractions and understand the guest motivation to visit them. Prepare a note on essential visitor segments and their motivation. For example, which one of these segments do you believe are underserved?
3. Prepare a business plan for your tour guiding company.
4. Identify a theme for interpretation and develop a sample itinerary.
5. What are the fundamental principles of designing an itinerary?
6. Who are essential partners for a tour guide? Who do you think are the two most important ones for you?
7. Why should a tour guide be aware of the legal framework?
8. Is tour guiding legislated in your country? What are some of the salient features of this legislation?
9. Assess the local heritage and environmental laws and evaluate how they affect visitors and local tourism.
10. Health laws have become significant in the wake of the Covid-19 pandemic. How do you think will they affect tour guiding in the short and long run?
11. How effectively do you think codes of conduct extract ethical behaviour from tour guides? Which practices in local tour guiding would you consider not so ethical?

Activity

1. Visit a monument and design two itineraries. One detailed and other around 50% of the duration.
2. Visit the local authorities and figure out the process of registering a guiding company

References

Chowdhary, N. (2014). *Handbook for Tour Guides*. New Delhi: Matrix Publishers.

Crowley, L. G., & Karim, M. A. (1995). Conceptual model of partnering. *Journal of Management in Engineering, 11*(5), 33-39.

Gauteng Guides Association (2000). *The Tourist Guide Code of Conduct and Ethics*. Retrieved from: https://www.guidessa.org/sites/default/files/2019-02/The%20 Tourist%20Guide%20Code%20of%20Conduct%20and%20Ethics.pdf

Gracyk, T. (2012, Feb 3). *Four fundamental ethical principles*. Retrieved April 6, 2021, from http://web.mnstate.edu/gracyk/courses/phil%20115/Four_Basic_principles.htm

Greenwood, V. A., & Dwyer, L. (2015). Consumer protection legislation: A neglected determinant of destination competitiveness? *Journal of Hospitality and Tourism Management, 24*, 1-8. doi:http://dx.doi.org/10.1016/j.jhtm.2015.04.001

IEP (1995). *Ethics.* Retrieved April 6, 2021, from *Internet Encyclopedia of Philosophy*: https://iep.utm.edu/ethics/

Mizoram Tourism (2020). *Guidelines for Tour Guides in Mizoram.* Tourism Department, Government of Mizoram. Retrieved April 7, 2021, from: https://tourism.mizoram.gov.in/upload/pages/Guidelines%20for%20Tour%20Guides.pdf

OECD. (2020, June 2). *Tourism Policy Responses to the coronavirus (COVID-19).* Retrieved April 2, 2021, from https://www.oecd.org/coronavirus/policy-responses/tourism-policy-responses-to-the-coronavirus-covid-19-6466aa20/

Sarasvathy, S. D. (2001). Causation and effectuation: Toward a theoretical shift from economic inevitability to entrepreneurial contingency. *Academy of Management Review, 26*(2), pp. 243-263.

Sullivan, A. M. (2016). Cultural Heritage & New Media: A Future for the Past. *The John Marshall Review of Intellectual Property Law, 604.*

The Guild. (2017). *Guide Code of Ethics.* Retrieved April 7, 2021, from Guild of Professional Tour Guides of Washington D.C.: https://www.washingtondctourguides.com/page/ethics

TIC. (2010). *Code of Conduct for the Tourist Guides.* Retrieved April 7, 2021, from Travel Industry Council of HongKong: http://www.tichk.org/public/website/en/guides/code_of_conduct_e_2010_07.pdf

Velasquez, M., Andre, C., Shanks, T. S., & Meyer, M. J. (2010). *What is Ethics?* Retrieved April 6, 2021, from Markkula Center for Applied Ethics: https://www.scu.edu/ethics/ethics-resources/ethical-decision-making/what-is-ethics/

WFTGA. (2006,). *Code of Guiding Ethics.* Retrieved from World Federation of Tourist Guide Associations: http://wftga.org/page.asp?id=16

Index